Springer Series in

MATERIALS SCIENCE 117

Springer Series in
MATERIALS SCIENCE

The Springer Series in Materials Science covers the complete spectrum of materials physics, including fundamental principles, physical properties, materials theory and design. Recognizing the increasing importance of materials science in future device technologies, the book titles in this series reflect the state-of-the-art in understanding and controlling the structure and properties of all important classes of materials.

99 **Self-Organized Morphology in Nanostructured Materials**
Editors: K. Al-Shamery and J. Parisi

100 **Self Healing Materials**
An Alternative Approach to 20 Centuries of Materials Science
Editor: S. van der Zwaag

101 **New Organic Nanostructures for Next Generation Devices**
Editors: K. Al-Shamery, H.-G. Rubahn, and H. Sitter

102 **Photonic Crystal Fibers**
Properties and Applications
By F. Poli, A. Cucinotta, and S. Selleri

103 **Polarons in Advanced Materials**
Editor: A.S. Alexandrov

104 **Transparent Conductive Zinc Oxide**
Basics and Applications in Thin Film Solar Cells
Editors: K. Ellmer, A. Klein, and B. Rech

105 **Dilute III-V Nitride Semiconductors and Material Systems**
Physics and Technology
Editor: A. Erol

106 **Into The Nano Era**
Moore's Law Beyond Planar Silicon CMOS
Editor: H.R. Huff

107 **Organic Semiconductors in Sensor Applications**
Editors: D.A. Bernards, R.M. Ownes, and G.G. Malliaras

108 **Evolution of Thin-Film Morphology**
Modeling and Simulations
By M. Pelliccione and T.-M. Lu

109 **Reactive Sputter Deposition**
Editors: D. Depla and S. Mahieu

110 **The Physics of Organic Superconductors and Conductors**
Editor: A. Lebed

111 **Molecular Catalysts for Energy Conversion**
Editors: T. Okada and M. Kaneko

112 **Atomistic and Continuum Modeling of Nanocrystalline Materials**
Deformation Mechanisms and Scale Transition
By M. Cherkaoui and L. Capolungo

113 **Crystallography and the World of Symmetry**
By S.K. Chatterjee

114 **Piezoelectricity**
Evolution and Future of a Technology
Editors: W. Heywang, K. Lubitz, and W. Wersing

115 **Defects, Photorefraction and Ferroelectric Switching in Lithium Niobate**
By T. Volk and M. Wöhlecke

116 **Einstein Relation in Compound Semiconductors and Their Nanostructures**
By K.P. Ghatak, S. Bhattacharya, and D. De

117 **From Bulk to Nano**
The Many Sides of Magnetism
By C.-G. Stefanita

Volumes 50–98 are listed at the end of the book.

Carmen-Gabriela Stefanita

From Bulk to Nano

The Many Sides of Magnetism

With 53 Figures

Springer

Dr. Carmen-Gabriela Stefanita
NanoDotTek
Burlington, MA 01803, USA
E-mail: info@nanodottek.com

Series Editors:

Professor Robert Hull
University of Virginia
Dept. of Materials Science and Engineering
Thornton Hall
Charlottesville, VA 22903-2442, USA

Professor R. M. Osgood, Jr.
Microelectronics Science Laboratory
Department of Electrical Engineering
Columbia University
Seeley W. Mudd Building
New York, NY 10027, USA

Professor Jürgen Parisi
Universität Oldenburg, Fachbereich Physik
Abt. Energie- und Halbleiterforschung
Carl-von-Ossietzky-Strasse 9–11
26129 Oldenburg, Germany

Professor Hans Warlimont
Institut für Festkörper-
und Werkstofforschung,
Helmholtzstrasse 20
01069 Dresden, Germany

Springer Series in Materials Science ISSN 0933-033X

ISBN 978-3-540-70547-5 e-ISBN 978-3-540-70548-2

Library of Congress Control Number: 2008931053

Typesetting: Data prepared by SPi using a Springer TEX macro package
Cover concept: eStudio Calamar Steinen
Cover production: WMX Design GmbH, Heidelberg

SPIN: 12255337 57/3180/SPi
Printed on acid-free paper

9 8 7 6 5 4 3 2 1

springer.com

In memory of my grandparents,
whose lifelong dedication made it all possible

Preface

The inspiration for this book can be traced back many years to two major works that influenced the author's outlook on applied physics: *Ferromagnetismus* by R. Becker, W. Döring (Springer, Berlin 1939), and *Ferromagnetism* by R.M. Bozorth (IEEE Press, New York 1951). The former work is a collection of lectures held in the 1930s for 'technicians' attending a technical college. The German language in which the work was originally written was extremely convenient for the author of this present book, as it was for a long time the only comfortable technical language in an English speaking environment. Later on, upon encountering the work by Bozorth, it was a relief to see the clarity and eloquence of the subjects presented in English, despite the impressive thickness of the book. Bozorth's work still constitutes a practical review for anyone in a multidisciplinary industry who comes across the various manifestations of magnetism. The popularity of both works is so enduring that they are regarded as highly academic, and yet extremely readable, a reference in their own right, still attracting many readers these days in industry and academia.

The field of magnetism progressed immensely in the twentieth century, and shows no signs of slowing down in the present one. It has become so vast that it is quite often viewed only in its parts, rather than as a whole. In today's myriad of applications, especially on a nanoscale, and their changeable implications mostly on a macroscale, it often seems that different aspects of reported work on magnetism are scattered and unrelated. Furthermore, the many atomic theories found in all major books on magnetism employ complex mathematical language that makes it less obvious how a theoretical description involving, e.g. spin can be associated with actual experimental observations.

The diverse expressions of magnetic phenomena on more than one scale, and the apparent confusion created by the overwhelming literature that treats disparate accounts of magnetism individually without placing them in a broader context, have led to the writing of this book. Based on the author's own struggle and experience in sifting through and organizing the vast amount of information, this work addresses the relationship between individual topics in magnetism, trying to make the connection between magnetic phenomena on various scales more understandable. Nevertheless, the author makes no claims that the book comes even close to the work of the masters mentioned earlier. The intention of this author is only to show how the different sides of magnetism come together. For this reason, the focus of the book is only on a few selected topics that the author believes are more representative of the broader subject.

The book has an introductory chapter on some basic concepts in magnetism. A few of these are later 'picked up' in subsequent chapters, while others are not mentioned again. Nevertheless, just highlighting them once draws the reader's attention to their existence and hints of their usefulness. The second chapter is an underpinning of magnetic nondestructive techniques, in particular magnetic Barkhausen noise, regarded by many as merely a laboratory nondestructive evaluation method. In any case, the valuable results and understanding gained through it have proved useful to more industrial nondestructive techniques such as Magnetic Flux Leakage and Remote Field Eddy Current. In the third chapter, the author takes a closer look at combined phenomena with wide industrial applications. The simple fact that optics and magnetism or piezoelectricity and magnetostriction can coexist has amazing consequences in many multidisciplinary areas. Furthermore, these subjects may recur in other established fields of magnetism, as implied in subsequent chapters. The fourth chapter goes deeper into the origins of ferromagnetism, showing that these constitute the foundation of emerging semiconductor electronics spin-offs (Chap. 5), as well as the recording heads in our everyday computers. The controversial and yet extremely promising field of spintronics is briefly described in Chap. 5, while some trends in magnetic recording media are tackled in Chap. 6.

Magnetism is used across many disciplines because of its rich implications in physics, chemistry, biochemistry, and the various areas of engineering. The author has undertaken to illustrate the various subfields in magnetism in a manner that anyone with a basic familiarity with modern physics can follow, regardless of their specialty. By no means is this book intended to be a comprehensive inclusion of all aspects of magnetism, nor does it have any claims that it treats the various areas in an exhaustive manner. On the contrary, this work is primarily intended to link the different areas of magnetism by showing how various phenomena fit into a broader picture. Its goal is to bring together a broad field in such a way that it provides a starting point for a graduate student or an experienced researcher for tackling a complex issue with maximum efficiency.

Collecting many sides of magnetism into a single volume had to be unavoidably selective; it is just an attempt at trying to spark an interest in this extended subject while keeping it together. Sometimes, this work has attempted to clarify the nature of macroscopic magnetic phenomena and how, in some cases, they can be traced back to a nanoscale. These days, the popularity of nanotechnology may overshadow macro phenomena, although they are closely connected. Nanotechnology deals with the manipulation of materials on an atomic or molecular scale measured in billionths of a meter, while having manifestations on an every day scale. At other times, the spotlight of the book has been on explaining the physical nature of some basic magnetic phenomena, while illustrating the connection with real applications or contemporary research.

It is a pleasure to acknowledge the support and encouragement I have received from colleagues and friends without whom I may have never written this book. My thanks go to Profs. L. Clapham and D.L. Atherton, as well as Drs. J.-K. Yi and T. Krause who may have long have forgotten how it all started. More recently, Prof. S. Bandyopadhyay, and my collaborators Drs. M. Namkung, F. Yun, and S. Pramanik have left their intellectual imprint on this work, therefore my gratitude extends to them. I apologize to all those who have not been named. Rest assured your influence has played a tremendous role in shaping this book, and the many subjects tackled are a tribute to your work.

Lastly, it should be stated that the author does not endorse any of the commercial products discussed in this book. The products were only mentioned for historical reasons, or to illustrate a principle and explain some magnetics concepts.

Burlington, MA,
July 2008

Carmen-Gabriela Stefanita

Contents

Symbols

B_J	Brillouin function
bcc	Body centered cubic
c	Crystallographic axis
d	Diameter of first Airy ring nm
d	Displacement of the magnetic wall nm
$\mathrm{d}V$	Unit volume cm^3
e	Electron charge $1.602176487 \times 10^{-19}\mathrm{C}$
E	Induced electric field (intensity) $\mathrm{V\,cm^{-1}}$
E_{exchange}	Exchange energy (density) $\mathrm{J\,cm^{-3}}$
E_{m}	Eigenvalues of $\hat{H}$ (also known as magnetic energy) J
$E_{\mathrm{magnetocrystalline}}$	Magnetocrystalline (anisotropic) energy (density) $\mathrm{J\,cm^{-3}}$
$E_{\mathrm{magneotelastic}}$	Magnetoelastic energy (density) $\mathrm{J\,cm^{-3}}$
$E_{\mathrm{magnetostatic}}$	Magnetostatic energy (density) $\mathrm{J\,cm^{-3}}$
E_{wall}	Energy (density) per unit surface area and unit wall thickness $\mathrm{J\,cm^{-3}}$
$\boldsymbol{F}$	Magnetic force N
g	Electron spin $g-$factor
g	Landé g-factor
g_0	Free electron g-factor, $g_0 = 2.0023$
$G_\uparrow$	"Spin-up" conductance S
$G_\downarrow$	"Spin-down" conductance S
$G_\uparrow^{\mathrm{tot}}$	Total "spin-up" conductance S
$G_\downarrow^{\mathrm{tot}}$	Total "pin-down" conductance S
$\hbar$	Reduced Planck constant $\mathrm{J\,s}$ or $\mathrm{N\,m\,s}$
H	Applied magnetic field (intensity) $\mathrm{A\,m^{-1}}$ or Oe
H	External magnetic field (intensity) Oe
H	Magnetic field intensity Oe
$\hat{H}$	Hamiltonian
$\boldsymbol{H}_0$	Applied magnetic field intensity $\mathrm{A\,m^{-1}}$ or Oe
H_{cw}	Magnetic wall coercivity Oe

H_{ext}	External magnetic field (intensity) Oe
H_{M}	Demagnetizing field (intensity) Oe
H_{σ}	Magnetic field (intensity) due to Bloch wall energy gradient Oe
hcp	Hexagonal closed packed
I	Sensing current A
j_{e}	Net electric current (density) $\mathrm{A\,cm^{-3}}$
j_{M}	Net magnetization current (density) $\mathrm{A\,cm^{-3}}$
$\boldsymbol{J}$	Total atomic angular momentum units of $\hbar$
J_z	z component of J units of $\hbar$
$\langle J_z \rangle$	Expectation value of J_z units of $\hbar$
kT	Thermal activation energy J
l	Length of bar magnet cm
l_{sd}	Spin diffusion length nm
L	Torque N m
L	Minimum mark length nm
$\boldsymbol{L}$	Total orbital angular momentum units of $\hbar$
m	Magnetic moment $\mathrm{A\,m^2}$
m_J	Eigenvalues of J_z
$\boldsymbol{M}$	Magnetization $\mathrm{A\,m^{-1}}$ or Wb m; $1\,\mathrm{Wb\,m} = 1/4\pi \times 10^{10}$ gauss $\mathrm{cm^3}$
M	Maximum moment reached before unloading $\mathrm{kg\,m^2}$
M	Specimen magnetization T
M_{Y}	Moment of yield at outer surface $\mathrm{kg\,m^2}$
$\mathrm{MBN_{energy}}$	See text for description $\mathrm{mV^2\,s}$
MR	Magnetoresistance %
n	Density of states for majority ($\uparrow$) and minority ($\downarrow$) spin-polarized electrons $\mathrm{cm^{-3}}$or $\mathrm{J^{-1}}$
n	Number of electrons in an atom
n	Spin density at distance x from the interface $\mathrm{cm^{-3}}$ or $\mathrm{J^{-1}}$
n_0	Spin density at the interface $\mathrm{cm^{-3}}$ or $\mathrm{J^{-1}}$
N	Number of atoms in the sample (e.g. Avogadro number)
NA	Numerical aperture of the objective lens
NA	System numerical aperture
p	Direction of polarization (German: parallel)
p	Magnetic pole strength Wb
p	Pattern period nm
p	Recording wavelength for magnetic mark nm
P	Spin polarization of the ferromagnetic layer %
r	Distance between magnetic poles cm
$R(0)$	Resistance at zero magnetic field Ω
$R(H)$	Resistance at a magnetic field value H Ω
s	Direction of polarization (German: senkrecht)
S	Electron spin
$\boldsymbol{S}$	Total spin angular momentum units of $\hbar$

t	Sample thickness mm
T	Absolute temperature K
T_2	Transverse relaxation time ns
$T_\uparrow$	"Spin-up" transmission probability
$T_\downarrow$	"Spin-down" transmission probability
U	Potential energy J
v	Linear velocity of the rotating disk $\mathrm{m\,s^{-1}}$
$\mathrm{v_F}$	Fermi velocity $\mathrm{m\,s^{-1}}$
$\mathrm{v_w}$	Linear velocity of the magnetic wall $\mathrm{m\,s^{-1}}$
V	Detected voltage V
x	Distance from the interface nm
x	Domain wall position nm
x	Ratio of magnetic and thermal energies
X	Magnetic susceptibility $\mathrm{H\,m^{-1}}$
z	Axis in the x, y, z Cartesian system
Z	Partition function

Greeks

α	Energy (density) contribution responsible for an easy axis $\mathrm{J\,cm^{-3}}$
α	Magnetoelectric voltage coefficient $\mathrm{V\,cm^{-1}\,Oe^{-1}}$
α	Parameter measuring spin transport asymmetry
β	Energy (density) contribution from the isotropic background $\mathrm{J\,cm^{-3}}$
β	Bohr magneton $\frac{\lvert e\rvert\hbar}{2mc} = 10^{-24}\,\mathrm{J\,T^{-1}}$
η_M	Spin injection efficiency for a single heterojunction
η'_M	Spin injection efficiency for a double heterojunction
Θ_N	Néel temperature K
θ	Angle between the directions of magnetic field and bar magnet magnetization
θ	Angle at which a magnetic field is applied
θ	Half-angle between the two beams in interference lithography
κ_α	Magnetic anisotropy constant $\mathrm{J\,cm^{-3}}$
λ	Magnetostriction
λ	Light wavelength used in lithography nm
λ	Mean free path nm
λ	Readout light wavelength nm
λ_s	Isotropic saturation magnetostriction
λ_{100}	Saturation longitudinal magnetostriction along [100]
λ_{111}	Saturation longitudinal magnetostriction along [111]
μ_0	Permeability of vacuum $\mu_0 = 4\pi \times 10^{-7}\,\mathrm{H\,m^{-1}}$
μ_B	Bohr magneton $\mu_B = 9.27400949(80) \times 10^{-24}\,\mathrm{J\,T^{-1}}$
μ_J	Component of μ parallel to J Bohr magnetons
μ_{J_z}	Projection of $\boldsymbol{\mu}_J$ along z Bohr magnetons
μ_w	Wall mobility $\mathrm{cm^2\,V^{-1}\,s}$

$\langle \mu_{J_z} \rangle$	expectation value of the magnetic moment μ_{J_z} Bohr magnetons
μ_L	Magnetic moment associated with **L** Bohr magnetons
μ_S	Magnetic moment associated with **S** Bohr magnetons
μ	Total magnetic moment Bohr magnetons
ρ	Resistivity corresponding to "spin-down" and "spin-up" electrons $\Omega\,\mathrm{cm}$
τ	Wall displacement time ns
$\tau_{\uparrow\downarrow}$	Spin-flip time ns
υ	Magnetic switching volume cm^3
φ	Easy axis direction
χ_0	Relative susceptibility
χ	Magnetic susceptibility

1

Introduction

Summary. The reader is introduced to some historical concepts in magnetism discovered over half a century ago, but of significant usefulness nowadays. Some magnetic nondestructive testing techniques, as well as magnetic tapes are based on a few of these concepts. Furthermore, many modern applications of magnetism rely on atomic scale magnetic phenomena that reach macroscopic values even at a few nanometers. With the advent of nanotechnology and its widespread implications, these concepts are the foundation for understanding a few of those that rely on magnetic properties. Nevertheless, the first chapter does *not* discuss *all* magnetic concepts or magnetic phenomena on which applications such as nanomechanical devices, spin valves, or quantum computing are based. On the contrary, the whole purpose of this book is to gradually entice the reader to discover the many sides of this discipline termed *magnetism*. As the book progresses, more and more magnetic concepts are being revealed and placed in a contemporary application context. Most people are not aware that a significant number of modern conveniences are based on magnetic properties. Hence, the book aims at clarifying these facts.

Magnetism has stimulated the interest of humans for a few thousand years, offering the possibility for imaginative exploitation of magnetic properties. From the compass needle to magnetic storage media, the overwhelming variety of magnetic discoveries has covered a colossal range of applications [1]. Whether by incorporating naturally occurring magnetic materials, or fabricating advanced artificial magnetic structures, the human intellect has been tireless in the pursuit of novel technologies [2].

Many magnetic phenomena were discovered over half a century ago [1]. They are gaining recognition now because manipulating magnetic structures is leading to significant technological advancements, such as nanomechanical devices, spin valves [3], and quantum computing [4, 5]. Many people are not aware of the fact that a significant number of modern conveniences are based on magnetic phenomena [6]. Furthermore, many of the magnetic properties encountered, whether intrinsic or induced, have atomic origins, but become fulfledged on length scales of the order of a few nanometers [7]. From there, they become macroscopically observable [4]. This recurrent fact, explicit or implied, should be kept constantly in mind as it represents the foundation for the topics depicted throughout this book.

1.1 Review of Certain Historic Magnetic Concepts

The mineral called *magnetite* (Fe_3O_4), the first magnetic material discovered, takes its name after *Magnesia*, a region in Turkey. A pointed piece of magnetite turns approximately north–south if it is supported in air or on the surface of water [8]. Alternatively, a pivoted iron needle becomes magnetic if rubbed with magnetite, and hence positions itself north–south. The word *lodestone* is derived from this directional property of magnetite or magnetized iron, as it means, in old English, a stone that leads the way (or lode). However, not all magnetite can become lodestones. A certain composition and crystal structure are required, as well as a strong magnetic field such as the transient field produced by lightning. The beginnings of magnetism are covered in many books, among which Still's [8] or Guimarães' [9] offer a captivating account on the properties and history of lodestone, as well as other permanent magnets.

1.1.1 Magnetic Susceptibility

The most common way of classifying magnetic properties of materials is by their response to an applied magnetic field. Materials that are magnetized to a certain extent by a magnetic field are called *magnetic* [10]. In particular, it is the quantity termed *magnetic susceptibility* χ that characterizes the magnetic response through the relationship

$$\boldsymbol{M} = \chi \boldsymbol{H}_0, \tag{1.1}$$

where $\boldsymbol{M}$ is the magnetization, also known as the magnetic moment per unit volume, and $\boldsymbol{H}_0$ is the applied magnetic field intensity [11]. Magnetic susceptibility is usually a tensor and a function of both field $\boldsymbol{H}_0$ and magnetization $\boldsymbol{M}$. For a magnetically isotropic material, $\boldsymbol{M}$ is parallel to $\boldsymbol{H}_0$, and χ is reduced to a scalar quantity. The unit for the permeability of vacuum μ_0 is the same as for χ [12]. Hence, it is possible to measure χ in units of μ_0. In this case, the measured dimensionless quantity is called *relative susceptibility* and is denoted by χ_0

$$\chi_0 = \frac{\chi}{\mu_0}. \tag{1.2}$$

Values for relative susceptibilities range from 10^{-5} [12] (very weak) to 10^6 (very strong magnetism). In some cases, the relative susceptibility is negative. Or, the relationship between $\boldsymbol{M}$ and $\boldsymbol{H}$ is not linear, so that χ_0 depends on $\boldsymbol{H}$. The behavior of χ_0 leads to various types of magnetism [2]. The origins of magnetism can be traced back to the orbital motion and to the spin of electrons that obey the Pauli exclusion principle, which will be briefly reviewed in subsequent sections.

1.1.2 Classification of Magnetic Materials

Ferromagnetic materials contain spontaneously magnetized magnetic domains where an individual domain's magnetization is oriented differently with respect to the magnetization of neighboring domains [2]. The spontaneous domain magnetization is a result of unpaired electron spins from partially filled shells, spins aligned parallel to each other due to a strong exchange interaction. The arrangement of spins depends on temperature and so does the spontaneous domain magnetization [2]. When the total resultant magnetization for all magnetic domains is zero, the ferromagnetic material is said to be *demagnetized*. However, an applied magnetic field changes the total resultant magnetization from zero to a saturation value [2]. When the magnetic field is decreased and reverses in sign, the magnetization of a ferromagnetic material does not retrace its original path of values, the material exhibiting so-called *hysteresis* [2]. A strong ferromagnet exhibits a relative susceptibility of 10^6 [12].

In contrast to ferromagnetism, a weak form of magnetism termed *diamagnetism* is attributed mainly to the orbital motion of electrons viewed classically as a "current loop," creating a magnetic moment [11]. An external magnetic field induces a magnetic flux in the diamagnetic material which counters the change in the external field. Diamagnetic materials exhibit an antiparallel magnetization with respect to the direction of the applied magnetic field, opposing the latter according to Lenz's law. Thus, the magnetization of a diamagnetic material is proportional to the applied magnetic field as seen in Fig. 1.1. Diamagnets have a negative and very weak relative susceptibility, of the order 10^{-5} [12]. Consequently, if a few magnetic atoms exist in the material, their influence overshadows the diamagnetism. Nonmagnetic atoms may become spin polarized by neighboring ferromagnetic atoms.

Similar to ferromagnetism, *paramagnetism* is also attributed to unpaired electron spins. However, due to a different electron configuration, these spins are free to change their direction. Therefore, at certain temperatures they assume random orientations as a consequence of thermal agitation [11].

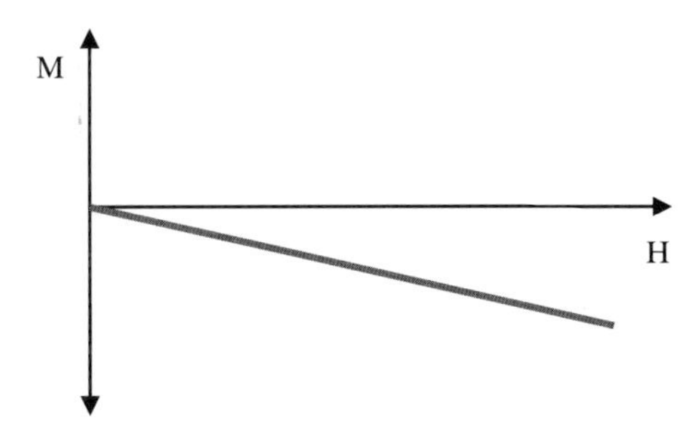

Fig. 1.1. Linear relationship between magnetization and applied magnetic field (intensity) in a diamagnetic material

An example of paramagnetism is the configuration of the electrons in the conduction band of metals [12]. When an external magnetic field is applied, a weak induced magnetization is produced parallel to the field. The induced magnetization is proportional to the external field, nevertheless stays positive, unlike in diamagnets. On the other hand, the susceptibility is inversely proportional to absolute temperature T, a fact also known as the *Curie–Weiss law* [11] (Fig. 1.2). For paramagnets, the relative susceptibility is a positive [12] 10^{-3} to 10^{-5}.

Analogous to paramagnetism, *antiferromagnetism* also exhibits a small positive relative susceptibility that varies with temperature [11]. However, this dependence differs significantly not only in the shape of the curve but also in the fact that in an antiferromagnetic material it displays a change at the so-called [11] *Néel temperature* Θ_N (Fig. 1.3). Below this temperature, the electron spins are arranged antiparallel so that they cancel each other and an external magnetic field is faced with a strong opposition due to the interaction between these spins. Consequently, the susceptibility decreases as

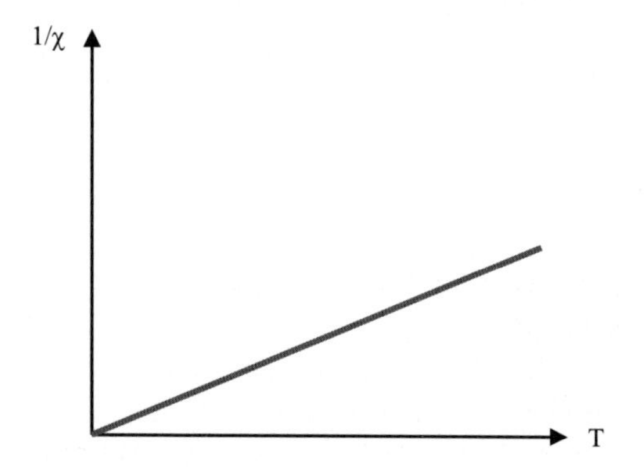

Fig. 1.2. Curie–Weiss law of paramagnetism, where the susceptibility is inversely proportional to absolute temperature

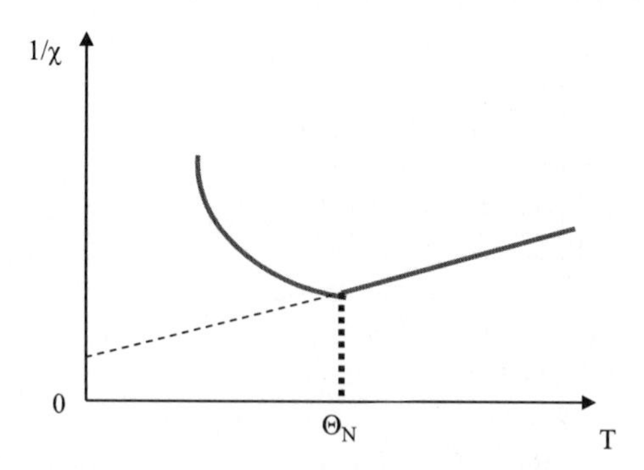

Fig. 1.3. Variation of susceptibility with temperature for an antiferromagnetic material

the temperature decreases, in contrast to paramagnetic behavior. However, above the Néel temperature the spins become randomly oriented while the susceptibility decreases as the temperature is raised [11].

Ferrites exhibit a kind of magnetism known as *ferrimagnetism*, in some ways similar to ferromagnetism [2]. However, in ferrimagnetic materials magnetic ions are placed on two different types of lattice sites, so that spins on one site type are oppositely oriented to spins on the other lattice site type [12]. The result is a total nonzero magnetization that is spontaneous. Nevertheless, an increase in temperature brings about a disturbance in the spin arrangement that culminates in completely random orientation of spins at the Curie temperature. At this temperature, the ferrimagnet loses its spontaneous magnetization and becomes paramagnetic. Ferromagnetic materials also have a Curie point above which they exhibit paramagnetic behavior [10, 13].

1.1.3 The Concept of Magnetic Pole

Quite often, the treatment of magnetism is similar to that of electrostatics [13, 14]. The fundamental magnetic phenomenon is viewed as an interaction between magnetic poles of strengths p_1 and p_2 separated by a distance r, analogous to the Coulomb interaction between electrically charged particles [12, 13]:

$$\boldsymbol{F} = \frac{p_1 p_2}{4\pi\mu_0 r^2}, \tag{1.3}$$

where $\boldsymbol{F}$ is the force acting on a magnetic pole and μ_0 is the permeability of vacuum mentioned above.

Alternatively, an electric current will also exert a force on a magnetic pole. Whether it is another magnetic pole or the electric current producing a magnetic field, the force $\boldsymbol{F}$ acting on a magnetic pole of strength p is [12, 13]

$$\boldsymbol{F} = p\boldsymbol{H}_0, \tag{1.4}$$

where $\boldsymbol{H}_0$ is the applied magnetic field. Magnetic poles occur in pairs. When a magnet is cut into pieces, each piece will have a pair of poles [11, 13].

Equation (1.4) implies that if a magnetic material is brought near a magnet, the magnetic field of the magnet will magnetize the material [12, 13]. Consequently, the magnetic field is sometimes called a *magnetizing force*. Furthermore, it is customary to represent the magnetic field by *lines*, also called *lines of force* (Fig. 1.4) to which a compass needle would be a tangent [11, 13]. As seen in Fig. 1.4, the magnetic field lines outside the magnet radiate outward from the north pole. They leave the north pole and return at the south pole, reentering the magnet [12].

If a bar magnet of length l which has magnetic poles p and $-p$ at its ends is placed in a uniform magnetic field, the couple of magnetic force gives rise to a torque [13] $\boldsymbol{L}$

$$\boldsymbol{L} = -pl\boldsymbol{H} \sin\theta, \tag{1.5}$$

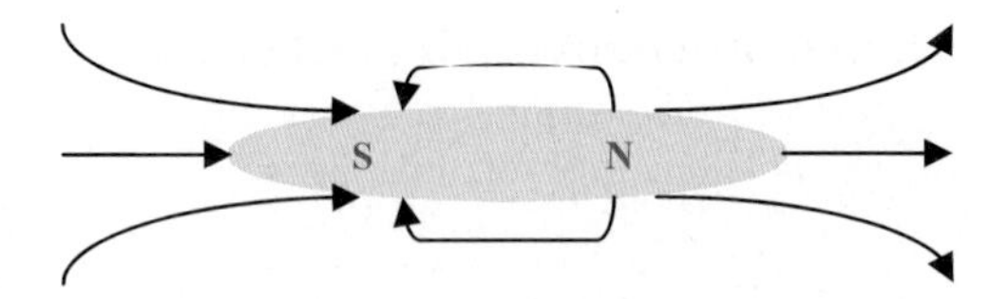

Fig. 1.4. Magnetic field representation outside a magnet or magnetized material

where θ is the angle between the direction of the magnetic field $\boldsymbol{H}$ and the direction of the magnetization $\boldsymbol{M}$ of the bar magnet [11]. The product pl is the magnetization $\boldsymbol{M}$ of the bar [13].

The work done by the torque gives rise to a potential energy U in the absence of frictional forces [11,13]

$$U = -\boldsymbol{M}\boldsymbol{H}\cos\theta. \tag{1.6}$$

This equation is particularly important in the discussion of magnetic domains and the realignment of their magnetization when an external magnetic field is applied [12]. The potential energy has a minimum value when [11,13] $\theta = 0$.

1.1.4 Magnetic Dipoles

If the bar length l tends to zero value, while simultaneously the strength of the magnetic poles p approaches infinity, the system thus produced is called a *magnetic dipole* [11,12]. Alternatively, a magnetic dipole can also be defined by a circular electric current of infinite intensity spanning an area of zero dimension [12,13]. No matter how we look at it, the magnetic dipole is only a mathematical concept, useful for the definition of some magnetic quantities.

The magnetic moment m of the magnetic dipole is [13]

$$m = \boldsymbol{M}\,\mathrm{d}V, \tag{1.7}$$

where $\boldsymbol{M}$ is the magnetization mentioned earlier, and $\mathrm{d}V$ is the unit volume. This equation was considered in earlier books as the definition [13] for $\boldsymbol{M}$. If the magnetization is constant throughout the magnetized body, the latter is considered homogeneous from a magnetic point of view [12].

1.2 Origins of Magnetism on an Atomic Scale

The magnetic moment of atoms originates from electrons in partly filled electron shells, and is determined by a fundamental property known as the *angular momentum* [15]. Each individual electron has an *angular momentum associated with its orbital motion*, and an intrinsic, or *spin angular momentum* [15]. Hence, there are two sources of the atomic magnetic moment: currents associated with the orbital motion of the electrons, and the electron spin [13].

1.2.1 The Importance of Angular Momentum

For an n-electron atom, these $2n$ angular momenta couple together to give a total angular momentum whose exact properties depend on the details of the coupling parameters [16]. The individual atomic orbital angular momenta couple together to give a total orbital angular momentum $\boldsymbol{L}$, and the individual atomic spin angular momenta couple together to give a total spin angular momentum $\boldsymbol{S}$. Finally, $\boldsymbol{L}$ and $\boldsymbol{S}$ couple together, to give a total atomic angular momentum [16] $\boldsymbol{J}$.

The orbital and spin angular momenta each have a magnetic moment associated with them

$$\boldsymbol{\mu}_L = -\beta \boldsymbol{L}, \quad \boldsymbol{\mu}_S = -2\beta \boldsymbol{S}, \tag{1.8}$$

where β is the *Bohr magneton.* The total magnetic moment $\boldsymbol{\mu}$ is then

$$\boldsymbol{\mu} = -\beta(\boldsymbol{L} + 2\boldsymbol{S}). \tag{1.9}$$

A system consisting of N identical magnetic atoms will have a total angular momentum $\boldsymbol{J}$ and magnetic moment $\boldsymbol{\mu}$. $\boldsymbol{L}$, $\boldsymbol{S}$, and $\boldsymbol{\mu}$ precess about $\boldsymbol{J}$. The component of $\boldsymbol{\mu}$ perpendicular to $\boldsymbol{J}$ averages to zero over a time significantly larger than the precession period [16]. When a field is applied, only the component of $\boldsymbol{\mu}$ parallel to $\boldsymbol{J}$ is sensed. That parallel component will be denoted $\boldsymbol{\mu}_J$.

The angular momentum state of an atom is characterized by eigenvalues of J [2], that is $J(J+1)$. Using the properties of angular momentum operators and the law of cosines, we have

$$\mu_J^2 = g^2 J(J+1)\beta^2. \tag{1.10}$$

Choosing the z component of $\boldsymbol{J}$, that is J_z with eigenvalues $m_j = J, J-1, \ldots, -J$, the magnetic moment along z is

$$\mu_{J_z} = -g\beta J_z, \tag{1.11}$$

where g, the Landé g-factor or spectroscopic splitting factor is given by*

$$g = 1 + \frac{J(J+1) + S(S+1) - L(L+1)}{2J(J+1)}. \tag{1.12}$$

Nevertheless, the Landé g-factor results from the calculation of the first-order perturbation of the energy of an atom when a weak external magnetic field acts on the sample [15,16]. Normally, the quantum states of electrons in atomic orbitals are degenerate in energy, thereby the degenerate states all share the same angular momentum. However, if the atom is placed in a weak magnetic field, the degeneracy is lifted [17]. Furthermore, this dimensionless g-factor relates the observed magnetic moment μ_{J_z} of an atom to the angular momentum quantum number m_j and the fundamental quantum unit of magnetism, that is the Bohr magneton [15,16].

* For a rigorous derivation of above results, please see any introduction to quantum mechanics [15, 16], or more specialized books on electric and magnetic susceptibilities [17].

1.2.2 Magnetic Moment of a Sample of N Atoms

In a simple paramagnet, the atoms do not interact with each other, and the only contributions to the Hamiltonian $\hat{H}$ come from their interaction with the applied magnetic field $\boldsymbol{H}_0$. As the atoms are identical, only the Hamiltonian for a single atom needs to be considered [15–17]

$$\hat{H} = -\boldsymbol{\mu}_J \cdot \boldsymbol{H}_0. \tag{1.13}$$

Choosing $\boldsymbol{H}_0$ to be along the z-axis, we can write [15, 16]

$$\hat{H} = -\mu_{J_z} \boldsymbol{H}_0. \tag{1.14}$$

The eigenvalues of H are then [15–17]

$$E_m = -g\beta m_j \boldsymbol{H}_0. \tag{1.15}$$

The partition function Z is defined by [15–17]

$$Z = \sum_n \exp\left(-E_m/kT\right) = \mathrm{Tr}\left(\exp\left(-\hat{H}/kT\right)\right) \tag{1.16}$$

or, in our case [15–17],

$$Z_J = \sum_{m_j=-J}^{J} \exp\left(g\beta m_j \boldsymbol{H}_0/kT\right). \tag{1.17}$$

The partition function is an important quantity when dealing with multi-particle structures, as it encompasses the statistical properties of the entire system [15–17]. It depends on a number of factors, such as the system's temperature, the angular momentum quantum number, external magnetic field, etc. Furthermore, it is a sum over all states while determining how the probabilities are divided among the various states composing the system, based on their individual energies [15–17].

The magnetic and the thermal energies can be expressed in terms of the partition function. Denoting by x the ratio of magnetic and thermal energies [15–17]

$$x = \frac{g\beta J \boldsymbol{H}_0}{kT}, \tag{1.18}$$

the partition function becomes

$$Z_J(x) = \sum_{m_j=-J}^{J} \exp(m_j x/J) = \frac{\sinh\left(\frac{2J+1}{2J}x\right)}{\sinh\left(\frac{1}{2J}x\right)}. \tag{1.19}$$

The partition function allows calculation of the expectation value of the magnetic moment μ_{J_z}, a quantity observed experimentally [15–17].

The magnetization of the sample of N atoms is given by [15, 16]

$$M = N \langle \mu_{J_z} \rangle = Ng\beta \langle J_z \rangle, \tag{1.20}$$

where

$$\langle J_z \rangle = \frac{\mathrm{Tr}\left(J_z \mathrm{e}^{-\hat{H}/kT}\right)}{\mathrm{Tr}\left(\mathrm{e}^{-\hat{H}/kT}\right)}. \tag{1.21}$$

Consequently,

$$M = Ng\beta \frac{\mathrm{Tr}\left[J_z \mathrm{e}^{\hat{H}/kT}\right]}{Z_J(x)}. \tag{1.22}$$

This expression can be reduced to the useful form [15, 16]

$$M = Ng\beta J B_J(x), \tag{1.23}$$

where B_J is called the *Brillouin function* [15, 16]. This function describes the dependency of the magnetization on the applied magnetic field, temperature, and the total angular momentum quantum number; hence it is a useful concept. It is used to derive important laws of magnetism, such as the Curie–Weiss law mentioned earlier [12].

1.2.3 Crystal Field vs. Spin–Orbit Coupling

The magnetic moment of atoms in magnetic materials, such as the iron-series transition-metal atoms in ferromagnetic metals (e.g., Fe, Co, Ni, YCo_5), and ferrimagnetic nonmetals (e.g., Fe_3O_4, NiO) is largely given by the spin, rather than orbital motion [10]. In this case, the spin moment $\boldsymbol{\mu}_S$ is equal to the number of unpaired electron spins. On the other hand, the orbital moment $\boldsymbol{\mu}_L$ is very small, typically of the order of $0.1\,\beta$, because the orbital motion of electrons is "quenched" by something called the *crystal field* [17].

Each atomic moment is acted on by the *crystal field*, proportional to the magnetization of its environment [12]. If an atomic moment were to be removed from its environment, it would leave behind a magnetic field. The field is produced by the surrounding spins, and is a manifestation of the local symmetry of the crystal. Crystal structure is a determining factor for intrinsic magnetic properties, such as saturation magnetization or magnetocrystalline anisotropy [2]. For example, the saturation magnetization of α-Fe (2.15 T) is associated with the bcc structure of elemental iron [13].

The competition between the electrostatic crystal field interaction and spin–orbit coupling is responsible for the alignment of atomic magnetic moments [13], giving rise to *magnetocrystalline anisotropy*.

1.2.4 Magnetocrystalline Anisotropy

Magnetocrystalline anisotropy is in effect a variation of magnetic properties with crystallographic orientation [2]. In iron, $\langle 100 \rangle$ is the preferred crystallographic direction along which magnetic moments from magnetic domains tend to align [2]. The anisotropy of most magnetic materials is of magnetocrystalline origin [13]. Permanent magnets, such as $SmCo_5$, or $Nd_2Fe_{14}B$ need a high magnetic anisotropy to keep the domain magnetization in a desired direction [13]. This is achieved due to the electronic configuration in these materials which results in a particular interaction between the crystal field and the spin–orbit coupling, as explained below.

The crystal field acts on the orbits of the inner shell *d* and *f* electrons. Concurrently, as a relativistic phenomenon spin–orbit coupling is most pronounced for inner-shell electrons in heavy elements, such as rare-earth *4f* electrons [13,15,16]. This results in a rigid coupling between spin and orbital moment in heavy elements [18]. On the other hand, the magnitude of the magnetocrystalline anisotropy depends on the ratio of crystal field energy and spin–orbit coupling [19].

It should be emphasized that for Fe, Ni, and Co, the magnetocrystalline anisotropy is due to *3d* electron spins, in contrast to the magnetocrystalline anisotropy for rare earths that originates in the *4f* shells [18]. In fact, the strong magnetocrystalline anisotropy in permanent magnets is given by the comparatively small electrostatic interaction of the unquenched *4f* charge clouds with the crystal field [13,17,18].

The absence of quenching means that typical single ion anisotropies (rare earth ions) are much larger than *3d* anisotropies [18,19]. This strong magnetocrystalline anisotropy is exploited in advanced permanent magnets, where it leads to very high coercivities, such as 4.4 T in $Sm_3Fe_{17}N_3$-based magnets [13,17,18]. Therefore, a large number of magnetic applications are based on rare earth metal alloys [18].

1.2.5 Magnetostriction

Aside from spontaneous magnetization and magnetocrystalline anisotropy, other intrinsic magnetic properties such as magnetostriction, or exchange stiffness also have origins in atomic scale magnetism [20]. Although they manifest themselves on length scales of a few angstroms, they reach bulk values even [11] at $\sim$1 nm.

Under the influence of a magnetic field, the shape of a ferromagnetic object changes due to a magnetic property termed *magnetostriction* [11] (noted λ). However, this type of deformation is very small, only of the order $\sim 10^{-5}$–10^{-6} [12], or even smaller in weakly magnetic materials. Magnetostriction was discovered in 1842 by Joule [11] who noticed a change in length when an iron rod was magnetized in a weak magnetic field, similar to the schematic illustration in Fig. 1.5.

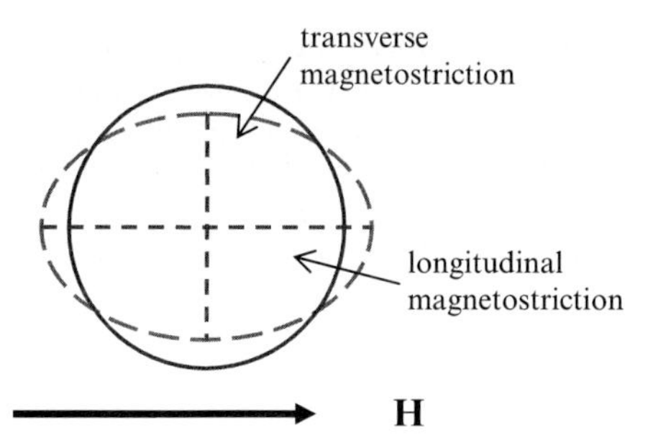

Fig. 1.5. Elongation of a ferromagnetic object in the direction of an applied magnetic field

Nevertheless, when a specimen elongates under an applied magnetic field, its volume remains constant. This means that a *transverse* magnetostriction exists, about half the value of the longitudinal magnetostriction, and of opposite sign [11].

Magnetostriction is believed to be due to spin–orbit coupling of valence electrons in ferromagnets [12]. Because electron orbits are coupled to spins, when the latter change direction to align with domain magnetization, the orbits change shape to conserve angular momentum. Since electron orbits are coupled to the crystal lattice, the lattice inside a *magnetic domain* (see below) deforms spontaneously in the direction of domain magnetization [12]. Iron single crystals magnetized to saturation in the [100] direction, increase in that direction due to magnetostriction.

The strain due to magnetostriction increases with magnetic field, until it reaches a saturation value. This value can be positive, negative, or in some alloys, zero [11]. Furthermore, magnetostriction saturates along a specific crystallographic axis, for example, $\lambda_{100} = 19.5 \times 10^{-6}$ and $\lambda_{111} = -18.8 \times 10^{-6}$ in a single cubic crystal. λ_{100} and λ_{111} are the saturation values of the longitudinal magnetostriction in the directions [100] and [111], respectively [21]. Quite often, an “isotropic saturation magnetostriction” $\lambda_s = \lambda_{100} = \lambda_{111} = -7 \times 10^{-6}$ is assumed [21], although it is not representative of experimental results [22].

1.3 Structure-Dependent Micromagnetism

Micromagnetic properties are usually structure dependent, and therefore responsible for a quite unique behavior of ferromagnetic materials under an applied magnetic field [2]. Some nondestructive evaluation techniques exploit micromagnetic properties to detect flaws and strains on the surface of engineering components [1]. Several aspects of nondestructive evaluation based on structure-dependent micromagnetism are discussed in more detail in Chap. 2.

Nevertheless, its basis is the fact that strongly magnetic materials divide spontaneously into magnetic domains [13] as a consequence of the minimization contest of five different energies, a process described briefly below.

1.3.1 Division into Magnetic Domains

In ferromagnetic materials, individual atomic magnetic moments tend to stay parallel to one another, keeping the exchange energy at a low value [13]. Such an alignment increases the magnetostatic energy by creating a large external magnetic field [11], as shown in Fig. 1.6. Therefore within the material, several magnetic domains are created, where within each domain individual magnetic moments add up to a total domain magnetization [1].

Furthermore, the domain magnetizations of neighboring domains are antiparallel [11] (Fig. 1.7). In this configuration, the exchange energy is somewhat increased, however the magnetostatic energy is lowered [11]. Domain walls are formed between magnetic domains [1].

1.3.2 Formation of Domain Walls

Further division into magnetic domains decreases magnetostatic energy even more, however the domain wall formed between domains with antiparallel magnetizations (Fig. 1.7) introduces an energy associated with the wall [11].

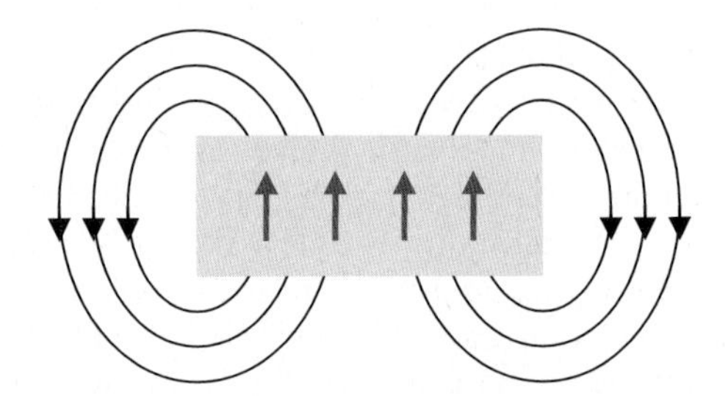

Fig. 1.6. Alignment of individual atomic moments increases magnetostatic energy by creating a large external magnetic field

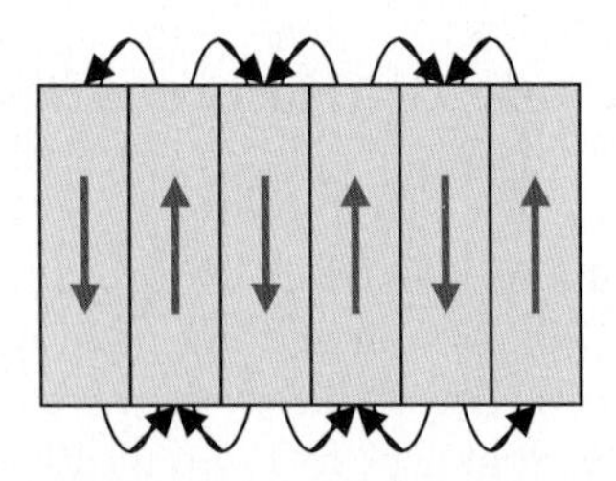

Fig. 1.7. Division into magnetic domains with antiparallel domain magnetizations decreases magnetostatic energy. A domain wall is formed between domains

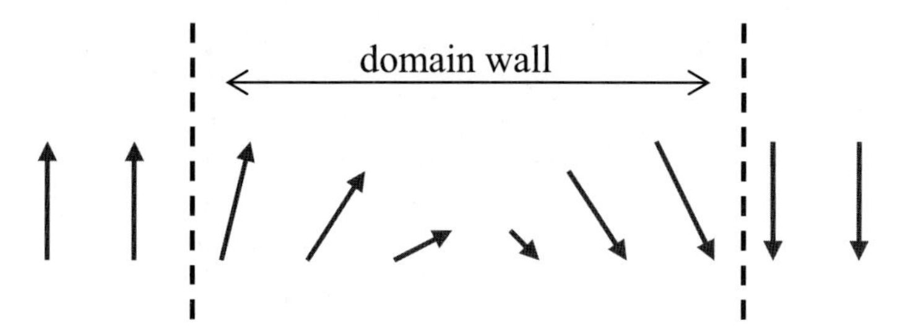

Fig. 1.8. Magnetic domain wall containing atomic magnetic moments of gradually varying orientation, ensuring a smoother transition to opposite domain magnetization

This wall increases the exchange energy which is highest at the wall [11]. Fortunately, the exchange force acts only over one or two atomic distances, having larger values in the wall vicinity.

If the transition from one magnetization direction to another is sharp, as is the case with antiparallel domain magnetizations, the exchange energy will be too high to keep this domain configuration in equilibrium [11]. Exchange energy [23] arises from the Pauli exclusion principle, and is a quantum-mechanical effect based on the degree of wavefunction overlap [15, 16].

A domain wall of a certain width, encompassing atomic magnetic moments of gradually varying orientation (Fig. 1.8), ensures a smoother transition opposite to domain magnetization direction, decreasing the exchange energy [11]. The width of the transition layer is determined, and thereby limited by the magnetocrystalline energy, which in order to maintain a minimum, tends to keep atomic magnetic moments aligned along one of the easy directions of the crystal axes [11].

1.3.3 Types of Domain Walls

The transition layer known as a domain wall can be of two types: *Bloch wall* [24] where the atomic magnetic moments rotate *outside* of the plane of the magnetic moments, and *Néel wall* [25] where atomic moments remain *in plane* while the rotation occurs [11]. Since domain magnetizations tend to align with preferred crystallographic axes, domain walls separating domains of different orientations can be classified as 180°, 90° (iron) or 109°, 71° (nickel), depending on the angles these crystallographic axes make in a specific lattice [11].

It should be noted that some of these walls of different orientation occur in *closure domains* [13]. The latter are created when the material divides into magnetic domains to allow more of the magnetic flux to stay within the material, minimizing magnetostatic energy [11, 13] (Fig. 1.9).

To monitor magnetostatic fields and domain configurations, colloidal suspensions termed [26] *ferrofluids* are usually employed. Ferrofluids, for instance

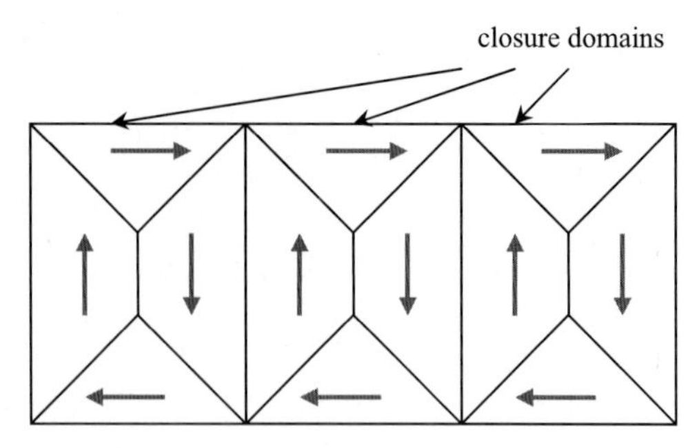

Fig. 1.9. Formation of 90° closure domains in iron. The closure domains are perpendicular to the 180° domains, illustrated here with vertical domain magnetizations

Fe_3O_4 or $BaFe_{12}O_{19}$ are stable substances, typically 10 nm particles immersed in hydrocarbons or other organic liquids, as water-based ferrofluids are more difficult to produce. They can also be used as liquids in bearings.

1.3.4 Significance of Magnetic Domains and Domain Walls

Magnetic domains and their walls are responsible for extrinsic magnetic properties, such as *remanence* and *coercivity.* They are also the reason for the *hysteresis* observed in ferromagnetic materials [27]. Magnetic domain configurations change with an applied magnetic field or stress through the displacement of domain walls. Therefore magnetic domains and in particular domain wall pinning by obstacles are magnetic microstructures exploited in some technological applications based on magnetic Barkhausen noise, or thickness dependent domain wall and coercive phenomena in thin films. On the other hand, in some cases such as magnetic recording, energetic losses created in the material because of these microstructures need to be minimized [28].

Particle size determines domain configurations and the mechanism of magnetization reversal within magnetic domains [29]. For instance, clusters are single-domain magnets and their large surface-to-volume ratio leads to strong diameter dependence of intrinsic properties such as anisotropy and magnetization [30]. Clusters tend to be *superparamagnetic*, particularly at high temperatures [31]. These aspects are important in magnetic recording.

Thin-film magnetism was initially also related to micromagnetic structures such as domains and domain walls [32]. Nowadays, it has developed into a separate branch of condensed matter physics [32] because nanostructured thin-films with intermediate or high coercivities are used in permanent magnets or magnetic recording [33], both strong and independent areas of magnetism.

1.4 Towards Technological Advancements

Magnetism is at present a very diversified discipline. Many magnetic structures, whether naturally occurring or artificially created have opened new possibilities for scientific and technological developments [34]. However, improving performance of existing magnetic materials is merely one of the many challenges of contemporary research [35]. Ultimately, new applications become feasible pushing not only nature, but also the human imagination, to its unexplored boundaries. A few of these topics will be discussed in subsequent chapters.

1.4.1 Design of New Magnetic Materials

Apart from natural magnets, most magnetic materials are not really known to the public at large. If faced with the option of drawing up a list of materials exhibiting magnetic properties, very few people can name more than three. To complicate matters even further, a new "materials design" approach is implemented these days in a few areas of magnetism.

One method is to nanoengineer structures to the extent that entirely new materials are fabricated. This creates artificial "metamaterials" relying on combined magnetic phenomena not observable in the original individual compounds. For instance, by introducing soft phases into hard ones, or incorporating a soft phase into an amorphous matrix, a completely different magnetic material can be created [36]. It is no wonder that we read reports of embedded magnetic clusters, or granular polymer materials displaying shape anisotropy [37].

Whatever the case may be, it shows that magnetic properties are realized on comparatively small length scales [38]. It also justifies why multiphase structures imitate the coexistence of independent magnetic properties [32]. Nevertheless, interesting magnetic systems have been produced by mechanical alloying and chemical reactions, with many possible applications: advanced magnetic recording media [39], materials for microwave applications, or even electroluminescent display devices [40].

1.4.2 Magnetic Quantum Dots

The search for ever-increasing storage densities in magnetic recording has led to the fabrication of two-dimensional arrays of nanodots. These are in essence very small structures where quantum-mechanical effects are no longer negligible, therefore coining the term *quantum dots* [41,42]. Concepts such as quantum well states and spin degrees of freedom come into play, expanding the areas of applications for quantum dots to quantum computing and spin electronics [43].

Several methods for producing quantum dots are currently being investigated, and some are reviewed later in this book. From more traditional ones, such as nanolithography, molecular beam epitaxy, or chemical vapor deposition, to emerging self-assembly techniques, whether DNA-assisted or chemically induced, complex arrays of quantum dots are presently being developed all over the world [4]. However, their long term survival will ultimately be determined by their ability to be implemented in large scale fabrication with minimum costs. Until then, these techniques are confined to the laboratory. In the following chapters, a closer look will also be taken at other types of nanostructures, and their feasibility examined.

The present book examines the fascinating realm of magnetic phenomena from a variety of angles, emphasizing current interests, as well as emerging trends in the ever-progressing technological miniaturization. Of course, the book is not all inclusive, however it opens the door for a sequence of topics that may be discussed in the future.

References

1. D.C. Jiles, *Introduction to Magnetism and Magnetic Materials* (Chapman and Hall, New York, 1991)
2. R.M. Bozorth, *Ferromagnetism* (IEEE Press, New York, 1951)
3. S. Mørup, C. Frandsen, Phys. Rev. Lett. **92**(21), 217201 (2004)
4. R. Skomski, J. Phys.: Condens. Matter **15**, R841 (2003)
5. E. Svoboda, IEEE Spectrum. 15 (2007)
6. R.B. Cowburn, A.O. Adeyeye, M.E. Welland, Phys. Rev. Lett. **81**(24), 5414 (1998)
7. J. Fidler, T. Schrefl, J. Phys. D: Appl. Phys. **33**, R135 (2000)
8. A. Still *Soul of Lodestone: The Background of Magnetical Science* (Murray Hill Books, New York, 1946)
9. A.P. Guimarães, *From Lodestone to Supermagnets: Understanding Magnetic Phenomena* (Wiley, New York, 2005)
10. M.E. Schabes, J. Magn. Magn. Mater. **95**, 249 (1991)
11. B.D. Cullity, *Introduction to Magnetic Materials*, 2nd edn. (Addison-Wesley, New York, 1972)
12. S. Chikazumi, *Physics of Magnetism* (Wiley, New York, 1964)
13. R. Becker, W. Döring, *Ferromagnetismus*, (Springer, Berlin, 1939)
14. F.T. Ulaby, *Fundamentals of Applied Electromagnetics* (Prentice Hall, New Jersey, 1999)
15. D.J. Griffiths, *Introduction to Quantum Mechanics* (Prentice Hall, New Jersey, 1995)
16. W. Greiner, *Quantum Mechanics, an Introduction*, 2nd edn. (Springer, Berlin, 1994)
17. Van Vleck, *Theory of Electric and Magnetic Susceptibilities* (Oxford University Press, Oxford, 1965)
18. J.M.D. Coey (ed.), *Rare-Earth Iron Permanent Magnets* (Oxford University Press, Oxford, 1996)
19. A. Aharoni, *Introduction to the Theory of Ferromagnetism* (Oxford University Press, Oxford, 1996)

20. J.S. Smart, *Effective Field Theories of Magnetism* (Saunders, Philadelphia, 1966)
21. C. Kittel, J.K. Galt, Solid State Phys. **3**, 437 (1956)
22. M.J. Sablik, G.L. Burkhardt, H. Kwun, D.C. Jiles, J. Appl. Phys. **63**(8), 3930 (1988)
23. W. Heisenberg, Z. Physik **49**, 619 (1928)
24. F. Bloch, Z. Physik **57**, 545 (1929)
25. L. Néel, Ann. Geophys. **5**, 99 (1949)
26. W.-L. Luo, S.R. Nagel, T.F. Rosenbaum, R.E. Rosenzweig, Phys. Rev. Lett. **67**, 2721 (1991)
27. W.F. Brown, *Micromagnetics* (Wiley, New York, 1963)
28. J.A.C. Bland, B. Heinrich (eds.), *Ultrathin Magnetic Structures*, vol. 1 (Springer, Berlin, 1994)
29. T. Schrefl, J. Fidler, K.J. Kirk, J.N. Chapman, J. Magn. Magn. Mater. **175**, 193 (1997)
30. S.E. Apsel, J.W. Emmert, J. Deng, L.A. Bloomfield, Phys. Rev. Lett. **76**, 1441 (1996)
31. C.P. Bean, J.D. Livingston, J. Appl. Phys. **30**, 120S (1959)
32. D.J. Sellmeyer, C.P. Luo, Y. Qiang, J.P. Liu, *Handbook of Thin Film Materials*, ed. by H.S. Nalwa, vol. 5 (Academic, San Diego, CA, 2002)
33. R.A. McCurrie, *Ferromagnetic Materials. Structure and Properties* (Academic, London, 1994)
34. R.M.H. New, R.F.W. Pease, R.L. White, J. Magn. Magn. Mater. **155**, 140 (1996)
35. R.W. Chantrell, D. Weller, T.J. Klemmer, S. Sun, E.E. Fullerton, J. Appl. Phys. **91**, 6866 (2002)
36. U. Gradmann, *Handbook of Magnetic Materials*, ed. by K.H.J. Buschow, vol. 7 (Elsevier, Amsterdam, 1993)
37. S.W. Charles, *Studies of Magnetic Properties of Fine Particles and Their Relevance to Materials Science*, ed. by J.L. Dormann, D. Fiorani (Elsevier, Amsterdam, 1992)
38. R. Skomski, J.M.D. Coey, *Permanent Magnetism* (Institute of Physics, Bristol, 1999)
39. I.A. Al-Omari, D.J. Sellmeyer, Phys. Rev. B **52**, 3441 (1995)
40. J.E. Evetts (ed.), *Concise Encyclopedia of Magnetic and Superconducting Materials* (Pergamon Press, Oxford, 1992)
41. A.V. Khaetskii, Y.V. Nazarov, Phys. Rev. B **64**, 125316 (2001)
42. H. Zeng, R. Skomski, L. Menon, Y. Liu, S. Bandyopadhyay, D.J. Sellmeyer, Phys. Rev. B **65**, 134426 (2002)
43. X.-D. Hu, S. Das Sarma, Phys. Rev. A. **61**, 062301 (2000)

2

Barkhausen Noise as a Magnetic Nondestructive Testing Technique

Summary. In a large part of the hysteresis cycle of a ferromagnetic material, the magnetization process takes place through a random sequence of discontinuous movements of magnetic domain walls, giving rise to what is termed *magnetic Barkhausen noise* (MBN). This noise phenomenon can give information on the interaction between domain walls and stress configurations, or compositional microstructure. It is also a complementary nondestructive testing technique to eddy-current probe sensors as well as magnetic flux leakage (MFL), both established in the nondestructive evaluation industry.

This chapter takes a closer look at the influence of stress on magnetic domain configuration, and how this is reflected in the MBN signal. The latter can be analyzed by using a variety of parameters, and some of these are introduced during the discussion. Apart from domain configuration, stress also affects magnetic anisotropy which can reveal further details on the stress state present in the material. Concurrently, residual stresses and dislocations play a significant role in the MBN investigation, completing the analysis and adding to the competitiveness of MBN as a nondestructive testing technique for ferromagnetic materials.

2.1 Introduction

In a large part of the hysteresis cycle of a ferromagnetic material, the magnetization process takes place through a random sequence of discontinuous movements of magnetic domain walls, giving rise to what is termed *magnetic Barkhausen noise* (MBN) [1,2]. This noise phenomenon is investigated statistically through the detection of the random voltage observed on a pick up coil during the magnetization of the material [3]. Analysis of MBN can give information on the interaction between domain walls and stress configurations, or compositional microstructure [4]. It is also a complementary nondestructive testing technique to eddy-current probe sensors [5–7] as well as magnetic flux leakage (MFL) [8], both more established in the nondestructive evaluation industry.

2.2 A Basic Definition of Magnetic Barkhausen Noise

As mentioned above, during the action of a smoothly varying alternating magnetic field of intermediate intensity, abrupt irreversible changes in the form of MBN emissions (Fig. 2.1) are observed in the magnetization of a ferromagnetic material [3]. These irreversible changes occur in the steep part of the magnetization curve, and they account for magnetic hysteresis in ferromagnetic materials [9]. MBN is named after its discoverer [10], and is called "noise" due to the sound heard in the loudspeaker used in the original experiment. It is termed "magnetic" to distinguish it from *acoustic Barkhausen noise*, the latter being based on magnetoacoustic emission [11,12].

2.2.1 Types of MBN Experiments

There are two types of Barkhausen noise experiments that are usually performed. If the detection coil is placed on the surface of the specimen, the emissions are termed *surface Barkhausen noise*, whereas a coil wrapped around the specimen detects *encircling Barkhausen noise* [13,14] (Fig. 2.2). According to

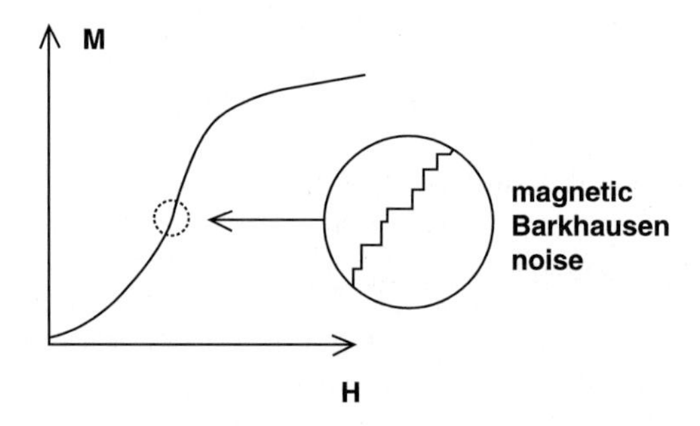

Fig. 2.1. Irreversible discontinuities in magnetization M as the ac magnetic field H is varied are termed magnetic Barkhausen noise

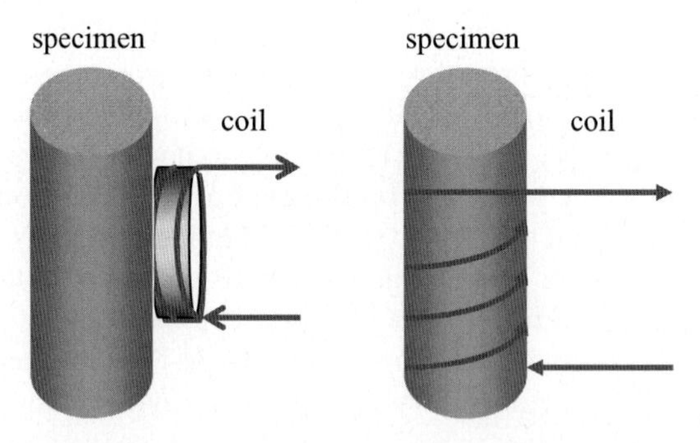

Fig. 2.2. Surface (*left*) vs. encircling (*right*) Barkhausen noise detection

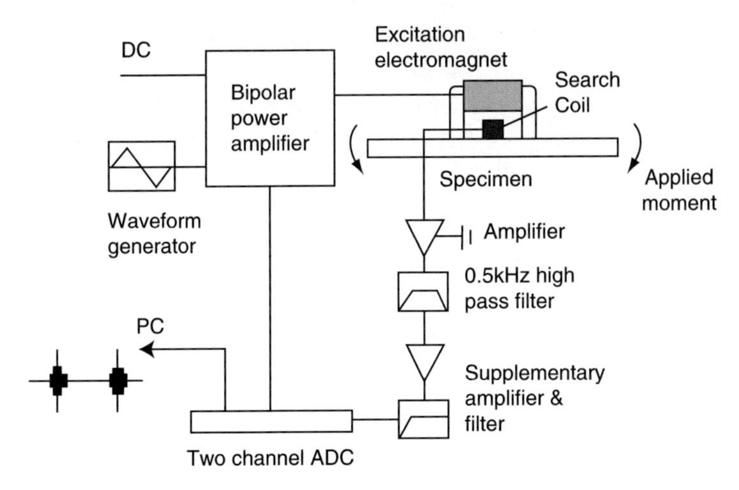

Fig. 2.3. A typical MBN measurement apparatus (reprinted from [16] (copyright 2004) with permission from Elsevier)

skin depth considerations [15], the estimated depth for minimum penetration of the magnetizing field is roughly 1 mm, whereas the depth from which the MBN signal originates is $\sim 30\,\mu\text{m}$. A typical MBN experimental setup [16] is sketched in Fig. 2.3. The MBN signal is detected by a search coil with a large number of turns of insulated copper wire wound around a ferrite cylinder. The output of the search coil is amplified and filtered [16].

2.2.2 Where does MBN Originate?

A ferromagnetic material that has not been magnetized consists of a large number of magnetic domains with random magnetic orientation, so that the bulk net magnetization is zero [3, 17] (Fig. 2.4). An external magnetic field tends to align the individual magnetic moments of the domains. Those domains with moments aligned most closely with the applied field will increase in volume at the expense of the other domains [9] (Fig. 2.5). The specimen becomes magnetized, as the walls move between adjacent domains [17].

When the external magnetic field is removed, the domains do not necessarily revert back to their original configuration [9]. This is because domain walls may have encountered pinning sites while moving, and to overcome these energy was expended [3, 9]. Once the wall has made it over the pinning site, there is no return path when the field is no longer acting. MBN is the irreversible "jump" of domain walls over local obstacles acting as pinning sites, such as grain boundaries, dislocations, inhomogeneities or other imperfections (Fig. 2.6). All lattice irregularities are likely to cause delays in domain wall movement, leading to uneven and discontinuous changes in magnetization [9, 18].

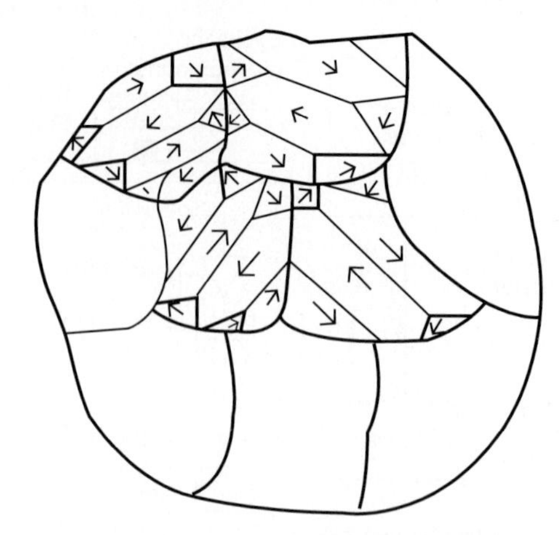

Fig. 2.4. Sketch of magnetic domains with random magnetic orientation in a polycrystalline ferromagnetic material, in the absence of an external magnetic field or stress. The *dark curves* represent grain boundaries (reprinted from [20])

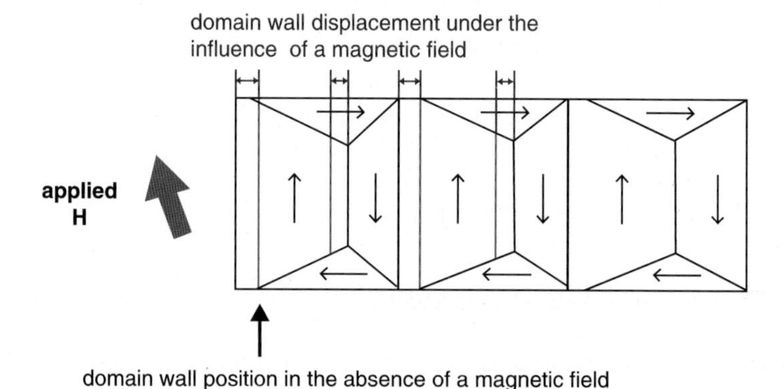

Fig. 2.5. Under the influence of an applied magnetic field, magnetic domains grow through wall displacement. Domains with moments aligned most closely with the applied field will increase in volume at the expense of the other domains. *Dashed lines* show wall positions in the absence of the field (reprinted from [20])

2.2.3 Formation of Magnetic Domains

Formation of magnetic domains occurs because of a minimization contest of the five basic energies involved in ferromagnetism:

$$E = E_{\text{exchange}} + E_{\text{magnetostatic}} + E_{\text{magnetocrystalline}} + E_{\text{magnetoelastic}} + E_{\text{wall}}. \tag{2.1}$$

The *exchange energy* E_{exchange} originates in quantum mechanical exchange forces or spin–spin interactions that are responsible for ferromagnetism [19]. A minimum in exchange energy is obtained when the spins of unpaired

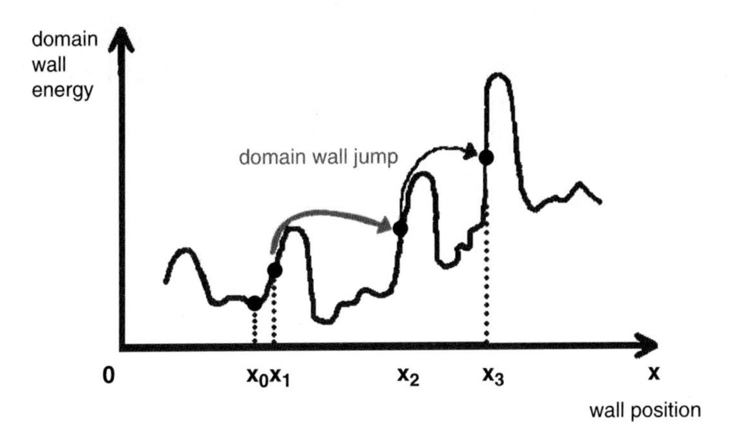

Fig. 2.6. Irreversible Barkhausen transitions. Domain walls overcome pinning sites and settle at energetically more favorable positions (reprinted from [20])

electrons are parallel, which is not possible in the same phase space. The *magnetostatic energy* $E_{\text{magnetostatic}}$ reaches a minimum when the magnetization of a magnetic domain is parallel to the external magnetic field [3, 9].

Crystal symmetry gives rise to a *magnetocrystalline (anisotropic) energy* $E_{\text{magnetocrystalline}}$ that becomes minimum when the magnetization of a magnetic domain is aligned with a preferred crystallographic direction, such as $\langle 100 \rangle$ in iron [15]. These directions are also termed *axes of easy magnetization* [3]. The crystal lattice strain is related to the direction of domain magnetization through the *magnetoelastic energy* $E_{\text{magneotelastic}}$ [9]. It is a minimum when the lattice is deformed such that the domain is elongated or contracted in the direction of domain magnetization [9].

The fifth energy is related to the fact that domain walls have certain energy per unit area of surface and unit thickness of wall E_{wall} because atomic moments are not parallel to each other, or to an easy axis.

Increases and decreases in these five energies have consequences for the equilibrium of the crystalline lattice in the material such that not all energies can be minimum at the same time. Formation of a certain magnetic domain configuration is the outcome of the *sum* of the five basic energies being minimized, although the energies themselves may not be at their minimum [20].

2.2.4 MBN and 180° Domain Walls

Domain walls separating regions of opposite magnetic moment are called 180° walls, whereas walls lying at 90° to each other are appropriately termed 90° walls [3, 15]. Nickel has 109° and 71° domain walls [3, 15].

It is believed that MBN is primarily due to 180° domain wall motion [3,9,21]. The 90° domain walls have stress fields associated with them, as their magnetizations lie at right angles on either side of the wall, causing lattice spacings to be slightly larger in the direction of magnetization. The resulting strain impedes with 90° domain wall motion, making them less competitive than 180° domain walls that have a higher velocity [22].

2.3 Stress Effects

Due to its high sensitivity to stress, MBN can be used as an independent nondestructive technique (NDT) for the evaluation of elastic stresses and strains [23], as well as plastic deformation and its consequences [20, 24].

2.3.1 Elastic Stress Causes Changes in Bulk Magnetization

A ferromagnetic material subjected to stress will cause changes to its bulk magnetization, even if no applied field is present [3]. This is because magnetic domains are influenced by stress and the resulting strain inside the material. Magnetic domains undergo stress-induced volume changes just like they would under an external magnetic field. As the internal elastic stress increases, the field required to move a domain wall across a pinning site, as well as the wall energy gradient increase, too [3]. The pinning sites themselves are also influenced by stress [25]. In fact, elastic strain effects are more influential on Barkhausen noise than plastic strain effects [25]. To gain a better idea of how stress influences MBN, a closer look at magnetic domains under the influence of stress is necessary.

2.3.2 Magnetic Domains Respond to Stress

An applied stress, just like an applied magnetic field, destroys the balance of the five energies. They have to readjust to minimize their sum [20]. If no external field is acting, the magnetostatic energy is zero. On the other hand, the magnetocrystalline and magnetoelastic energy are the dominant players [9, 15].

Under stress, both magnetocrystalline and magnetoelastic energy compete to determine the direction of the domain magnetization. Nevertheless, stresses in excess of 1,700 MPa are needed to counterbalance the effect of the magnetocrystalline energy, and change the direction of domain magnetization [26]. Therefore, the domain magnetization remains parallel to the *crystallographic* easy axis of the material, even under an applied stress.

Since realignment of magnetization with stress is not easily attained, the domain configuration minimizes its energy through movement of domain boundaries. A new energy configuration is achieved when domains lying closest to the direction of applied uniaxial tensile stress grow at the expense of domains with perpendicular domain magnetization. In contrast, under a compressive stress the domains with magnetic moments perpendicular to the axis of applied stress become energetically favorable [21]. In this manner, the magnetoelastic energy is decreased [15, 26].

There is another mechanism through which magnetic domains respond to an applied stress. In general under stress, there is an increase in the 180° domain wall population in the stress direction if the stress is tensile, with an opposite effect for compressive stress. Since MBN is associated with the pinning of magnetic domain walls, a higher signal is obtained along the direction of tensile stress, and a lower for compressive stress [27].

2.3.3 Magnetic Anisotropy and MBN

Magnetic properties of a material depend on the direction in which they are measured, and this phenomenon is known as *magnetic anisotropy* [3]. Only the magnetocrystalline anisotropy which is directly related to crystal symmetry is an intrinsic property of the material, while all other magnetic anisotropies are induced [15]. Quite often, design of commercial ferromagnetic materials is dependent on their magnetic anisotropy [23].

MBN is a suitable technique for detecting magnetic anisotropy in a ferromagnetic material, without regard to its origin. An MBN signal is detected at particular angles with respect to the specimen's axis, and after mathematical modeling it is plotted on a polar graph. The shape of the graph reveals the direction of the *magnetic easy axis* or magnetic anisotropy, because the MBN signal is large in that particular direction, as will be seen in subsequent sections.

2.3.4 Some Parameters Used in MBN Analysis

When modeling the MBN signal, two contributions are taken into account (1) that from domains responsible for an easy axis, also known as α and (2) a contribution from isotropically oriented domains represented by β [28]. The two contributions α and β are incorporated as fitting parameters into a mathematical expression describing a so-called $\mathrm{MBN}_{\mathrm{energy}}$

$$\mathrm{MBN}_{\mathrm{energy}} = \alpha \cos^2(\theta - \phi) + \beta, \qquad (2.2)$$

where θ is the angle at which a magnetic field is applied, and φ is the easy axis direction [28]. The parameter α is obtained by subtracting β from the maximum "$\mathrm{MBN}_{\mathrm{energy}}$," due to β being a minimum in $\mathrm{MBN}_{\mathrm{energy}}$. α represents contributions from domains responsible for a magnetic easy axis, whereas β takes into account contributions from isotropically oriented domains [20]. The significance of the $\mathrm{MBN}_{\mathrm{energy}}$ is related to the change in magnetic flux under the influence of the magnetic field component parallel to a particular domain wall [20]. The latter will move if its coercivity is overcome by the field component parallel to the wall [28].

MBN measurements are taken at 10° intervals covering the entire 360° circle of an angular scan. The angle is with respect to the sweep field direction of the MBN sensor [20]. Equation (2.2) is fit to the measured MBN data, while

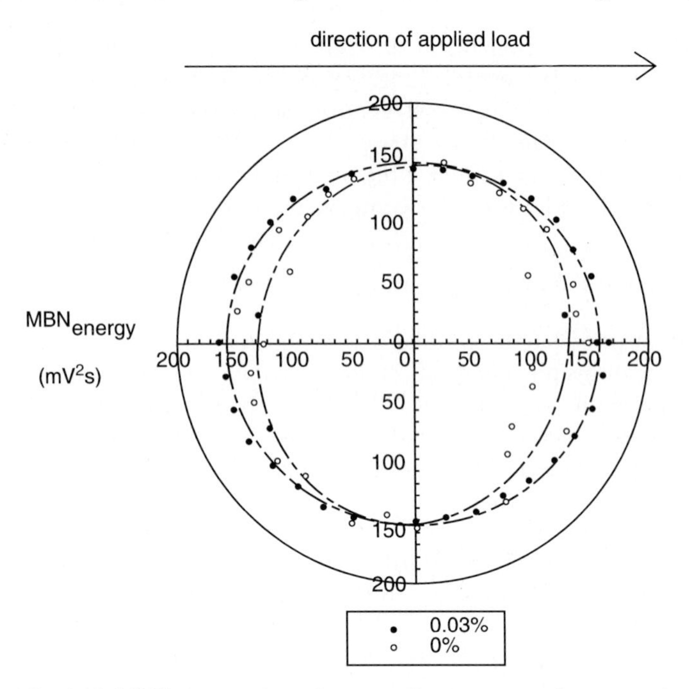

Fig. 2.7. Typical $\mathrm{MBN}_{\mathrm{energy}}$ polar plot revealing a magnetic easy axis (or magnetic anisotropy) under the influence of an applied uniaxial tensile load. The plot is elongated in the direction of magnetic anisotropy which in the case of the 0.03% deformation is also the direction of the applied uniaxial tensile stress. The plot at 0% deformation is not a perfect circle, indicating that a magnetic anisotropy is present in the as is material due to prior processing which resulted in trapped residual stresses. The magnetic anisotropy of the as is material is in a direction different from the one along which the load will be subsequently applied (reprinted from [41])

the resulting calculated $\mathrm{MBN}_{\mathrm{energy}}$ is plotted on a polar graph. Figure 2.7 shows a typical polar graph for the $\mathrm{MBN}_{\mathrm{energy}}$ when magnetic anisotropy is present. In the absence of magnetic anisotropy, the $\mathrm{MBN}_{\mathrm{energy}}$ plot is a perfect circle. If magnetic anisotropy is present such as for instance due to an applied uniaxial tensile load, the circle is elongated in the direction of the magnetic anisotropy which is also the direction of the uniaxial tensile load, transforming the plot from a circle into an ellipse [20].

Apart from $\mathrm{MBN}_{\mathrm{energy}}$, another parameter termed *pulse height distribution* is also used to characterize the MBN signal [20]. This is because the MBN signal consists of a collection of voltage pulses or "events" of varying amplitude that carry information about the magnetic state of the material. Usually, only voltage pulses above and below a certain threshold are considered in order to maintain integrity of the analysis. A positive slope between two consecutive measurements crossing the positive voltage threshold defines the onset of an event, while a similar positive slope crossing the negative voltage threshold the end of it.

The occurrence of events of different amplitude is represented in a graph termed a *pulse height distribution* where the absolute values of the event amplitudes are examined by height [20]. Actually, the MBN_{energy} is obtained by calculating for each event the area between the time axis and the squared voltage pulse, and summing over all measured events.

2.3.5 Elastic Stress Influences on Magnetic Anisotropy

Angular MBN scans are very sensitive to stress-induced changes; therefore, they are a good indicator of the stress condition present in a ferromagnetic material at the time of the investigation. A large number of studies have conclusively demonstrated a high magnetic response in steel to applied elastic tensile or compressive stress, with an easy axis development along the direction of the tensile (Table 2.1a), and away from the compressive stress direction [27,29–33].

Domain magnetization vectors are aligned with [100] crystallographic directions closest to the direction of principal stress [20]. An applied elastic tensile stress increases MBN_{energy} while creating a strong magnetic easy axis in the stress direction. The magnetic easy axis develops further with increased elastic stress, sometimes modifying its orientation while keeping up with directional variations in stress. Conversely, an applied compressive elastic stress decreases the MBN_{energy} in the compression direction [34], and creates a magnetic easy axis perpendicular to applied stress [29]. A magnetic easy axis becomes even more pronounced with plastic deformation, while MBN_{energy} values experience only a slight variation in the plastic regime, as opposed to the large increases observed during elastic deformation [25].

2.3.6 Plastic Deformation and Magnetic Anisotropy

A notable fact is that the magnetic easy axis continues to be present and become further pronounced in the plastic range of deformation [25] (Table 2.1c, e). Additionally, complex stress distributions left behind after mechanical processing such as cold rolling can also be characterized using MBN measured at different stages of rolling [20,35] (Table 2.1b, d, f). Table 2.1 gives an overview of what can be detected using MBN, simultaneously showing some of the influence of elastic and plastic deformation on magnetic domains. The latter are best represented by changes in pulse height distributions, especially noticeable along directions of maximum shearing stress [20] (Table 2.1c, left).

Plastic deformation distorts the crystalline lattice permanently by *slipping* when the critical-resolved shear stress is reached on a slip plane. Between slipped and unslipped portions dislocations form that alter the interplanar spacing, creating strain fields and thereby volume changes in magnetic domains. While dislocation strain fields contribute to the redistribution of strain within a grain, they also alter the magnetic texture of the material.

Table 2.1. Elastic vs. plastic deformation effects on $\mathrm{MBN}_{\mathrm{energy}}$, pulse height distribution, as well as magnetic domains (a) A magnetic easy axis develops under an elastic stress, becoming more pronounced in (c) and (e) under plastic deformation. Changes in pulse height distribution become apparent in (c, left) at an angle of 45° with respect to the principal stress, along the direction of maximum shearing stress. Cold rolling reveals a complex stress distribution in (b), (d), and (f) [35]

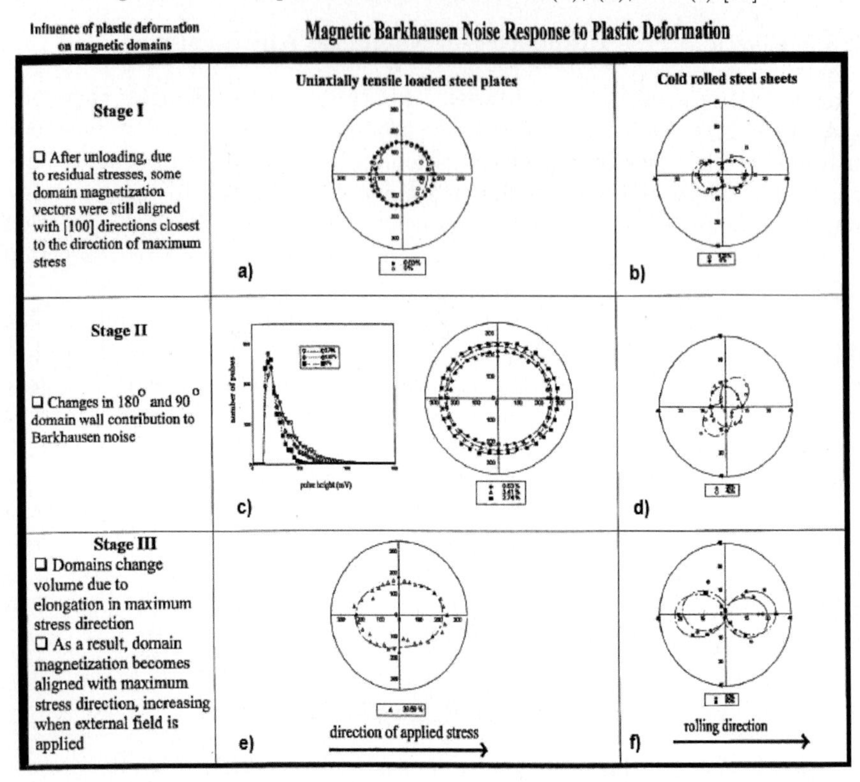

Therefore, it is no surprise that plastic stress-induced effects are reflected in the measured MBN signal which depends so strongly on changes in magnetic domain configuration [25]. While elastic strain has been noticed to significantly alter the magnetic anisotropy α in the specimen, it has little influence on the isotropic background signal β. On the other hand, plastic deformation has the opposite effect, in as much as it changes β, but leaves α almost unaltered [25].

2.3.7 Effects of Residual Stresses

When using angular MBN scans to characterize samples that have reached the plastic regime and have been unloaded, it is noticed that the easy axis becomes less pronounced, nevertheless is still present [35]. Furthermore, interestingly the curve described by (2.2) does not fit experimental data. Higher

MBN_{energy} values are revealed in both axial and transverse directions to the previously applied tension, indicative of residual stresses trapped by plastic strains [20]. Residual stresses in welded T-section samples of SAE 1020 steel have been successfully detected in the past [13]. Significant increases in MBN signal have been observed in areas surrounding the weld before the sample underwent stress relief treatment. The latter involves annealing, a process known to reduce magnetoelastic energy [36].

Residual stresses are elastic in nature; however, they can be surrounded by nonuniform microstructural changes in the specimen [16]. This is because grains within a polycrystalline sample deform differently depending on their crystallographic orientation with respect to the direction of applied stress. While deforming, grains can lock in elastic intergranular stresses [37], and their influence is visible in the MBN signal [13, 25]. Figure 2.8 shows a correlation

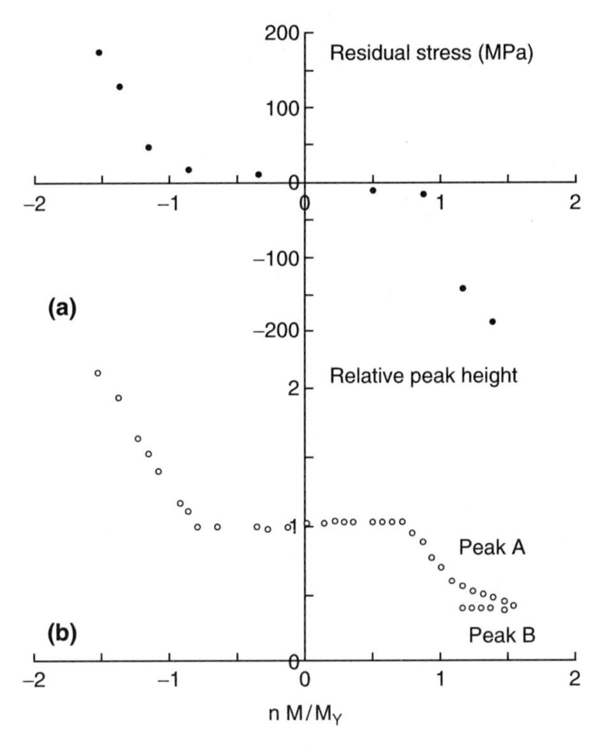

Fig. 2.8. Residual stress and relative MBN peak height in the unloaded state, correlated with the moment applied prior to unloading. (**a**) Residual stress determined by X-ray diffraction; (**b**) Normalized peak height from MBN profile. M/M_Y is the ratio between the maximum moment applied to a specimen before unloading, and the moment at which the stress reaches the yield point at the outer surface, while n is the sign of stress on the surface (positive for tension, and negative for compression) (reprinted from [16] (copyright 2004) with permission from Elsevier)

between residual stresses detected using X-ray diffraction and MBN peak height in unloaded specimens [16]. The same pattern is observed for both residual stresses and MBN peak height, proving that the two correlate very well.

2.3.8 Influence of Dislocations

Elastic stresses between grains adjust while dislocations appear in grain boundary regions [38], reducing stress concentration. Dislocations start to form even before the macroscopic elastic limit has been reached [39, 40]. The effect that dislocations have on the magnetic domain configuration can be witnessed through the abrupt and complex changes in $\text{MBN}_{\text{energy}}$ as well as pulse height distribution, changes characteristic of early stages of plastic deformation termed *microyielding* [40, 41]. The latter occurs as dislocation formation is initiated in some grains, while others remain intact [40].

The created strain fields result in increases in magnetic domains located in grain boundary regions. Their MBN signal is likely to be greater than the one of smaller size domains in the grain interior. A certain MBN response is obtained, dependent on how many grains develop dislocations. Not all grains undergo plastic deformation at the same time or at the same stress level. This is because of their different orientation with respect to the direction of applied stress. As deformation progresses reaching the *plastic stress regime*, massive dislocations are generated forming dislocation tangles as they move through the lattice. Also, more strain fields appear inside grains, as the bulk of the grain becomes affected by stress. New stress distributions emerge leaving their imprint on the MBN signal.

2.3.9 Selective Wall Energy Increases at Pinning Sites

Unequal size increases between magnetic domains, evidenced in the plastic stress regime, result in differences in the energetic level of domains. Furthermore, with enhanced dislocation density extending to more and more grains, domain wall energy gradient increases selectively at pinning sites [25]. Although single dislocations are too small to pin domain walls, dislocation tangles are believed to act as pinning sites to domain wall movement [42].

Figure 2.9 shows a possible domain wall energy redistribution at local pinning sites where walls experience a Barkhausen jump. When stress enters the plastic regime, some sites experience a significant energy increase, while simultaneously leaving other sites unchanged. Therefore, the Barkhausen transition is expected to occur from one high energy site to an even higher energy site, skipping sites with lower energy. Previous studies show that these unusual transitions are revealed in the detected MBN signal, in particular the pulse height distributions [20, 25].

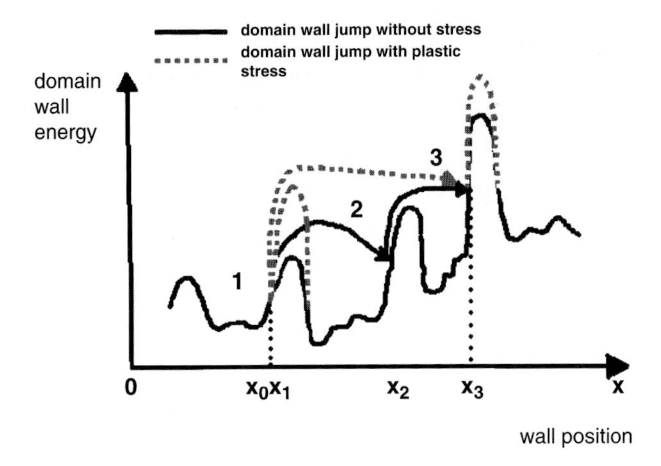

Fig. 2.9. Domain wall energy distribution at pinning sites with and without the influence of plastic stress. *Dashed lines* indicate changes that occur because of plastic stress, such as nonuniform increases in domain wall energy at some pinning sites and domain wall jumps that skip lower energy sites

2.3.10 Roll Magnetic Anisotropy

Variations in magnetic anisotropy are not only due to residual stresses or dislocations that alter magnetic domain configuration but also due to other factors such as crystallographic texture and various microstructural inhomogeneities [35,43]. These leave their imprint on MBN signals because they, too, determine domain wall dynamics.

MBN pulse height distributions are particularly indicative of such influences, as there are visible differences in their occurrence at various stages of plastic deformation. For instance, in the case of some typical nuclear reactor pressure vessel steel specimens subjected to varying levels of cold rolling, it was noticed that along the rolling direction the number of large MBN voltage pulses has diminished at reduction ratios of ∼25%, increasing again at ∼60% reduction ratios [35].

These pulse height distribution observations correlate well with the detected variations in the magnetic anisotropy of the nuclear reactor pressure vessel steel specimens, as the preexisting anisotropy direction was destroyed at ∼25% reduction ratio, increasing again and aligning with the rolling direction while deformation progressed to ∼60% reduction ratio [35]. It is known that a $\{100\}\langle 110\rangle$ texture develops in severely deformed steel, detectable after 20–30% deformation [44], therefore crystallographic texture development is one of the reasons magnetic anisotropy experiences changes.

However, there is significant scatter about the "ideal" texture until higher levels of deformation (80–90%) are reached when texture formation is complete [44]. Nevertheless, texture cannot be the only factor responsible for roll induced changes in magnetic anisotropy, at least judging by the direction of the latter displayed at different deformation levels. Stress levels are so high that inhomogeneous deformation is unavoidable, as internal elastic stresses become trapped in some parts of the material. These residual stresses, whether axially compressive due to the rolling process, or built up at grain boundaries or between crystallographic planes, bring their own contribution to the magnetic anisotropy of the cold rolled specimens.

The type of anisotropy described above is attributed to the cold rolling process itself, and has therefore been termed *roll magnetic anisotropy* [43,45,46]. It is believed to be due to complex metallurgical changes resulting in competing effects between crystallographic texture, as well as anisotropic microscopic and macroscopic residual stresses. In some cases, such as for nickel–cobalt alloys, a roll magnetic anisotropy has been observed along two directions, rather than a single one [46].

2.3.11 Limits in MBN Signal Increase with Plastic Stress

An applied stress that acts simultaneously on grains with different crystallographic orientations leads to longitudinal or shear incompatibility at grain boundaries [47]. The relative orientation of grains with respect to each other determines the type of incompatibility. Crystallographic orientations such as [100] and [111] experience longitudinal incompatibility [48], which means that one grain is in tension, while its neighbor is in compression. The stress at grain boundaries induced by longitudinal incompatibility is almost three times higher than the critical-resolved shear stress within a grain [49]. Therefore, dislocations are more likely generated at grain boundaries before they can form in the interior of the grain. Also, the interplanar spacing is altered in the vicinity of the actual dislocation, giving rise to what is termed *strain field of the dislocation* [47].

The formation of these dislocation induced strain fields results in local variations in interplanar distance along a certain crystallographic direction within a grain. Nevertheless, these can only accommodate so much elastic strain before slip systems are activated and plastic flow initiates, leading to work hardening. The latter will increase slightly the threshold for plastic flow, allowing some elastic strain to continue to build up, but this build up will eventually reach its limits. Mughrabi [50] advanced the idea of hard mechanical regions consisting of dislocation tangles in and near grain boundaries, surrounded by a softer matrix. The hard regions are built of small volumes assumed to be under tensile stress, while separated by larger volumes (soft regions) under compressive stress [50].

Thus, plastic deformation introduces permanent lattice distortion with different consequences for each particular domain configuration. Irregular stress distributions affect the volume of magnetic domains, especially for those in grain boundary regions that experience increases in volume before their neighbors in the grain interior. But every magnetic domain size modification implies wall movement, hence variations in MBN signal. Nevertheless, these volume changes can only extend so far, as the resulting additional strain fields attributed to massive plastic deformation will ultimately put restrictions in domain size increase, as well as impede the simultaneous motion of domain walls. These limitations lead to a slow down in MBN signal increase, fact evidenced in measurements performed at very high levels of deformation that show only a small rise in signal [41]. Some authors consider this effect a "magnetic degradation" [51,52].

2.4 Effects of Microstructure on MBN

2.4.1 Variations in Grain Size

Bertotti et al. [53] investigated the effect of grain size on MBN, observing that the boundaries of grains are likely sources for domain wall pinning. Ng et al. [54] confirmed that a large number of grain boundaries result in more intense Barkhausen noise emissions. The large number of boundaries is found in samples with smaller grains that have been annealed at lower temperatures. Small grain samples have larger MBN signals because the boundaries act as pinning sites, and since their fractional volume is larger, more pinning sites need to be overcome when the walls move.

Krause et al. [27] suggested that the number of 180° domain walls increases in the presence of applied tensile stress, and derived an expression for the change in magnetoelastic energy under these circumstances. From this expression, they calculated a threshold stress that would be necessary to add another domain wall to the configuration. This threshold stress depends on the grain size and it increases with the number of existing domain walls. Hence, MBN activity is strongly linked to grain size and grain boundaries.

Ng et al. [54] advanced the idea that the interaction between the domain walls and dislocation tangles leads to different MBN profiles than the interaction between the walls and grain boundaries. They used this argument to explain secondary peaks observed in some of the MBN signals. This is because the physical nature of the pin is assumed to dictate the restoring force acting on the wall.

Grain size influences the number of defects in the specimen and hence its magnetic properties [55]. Ranjan et al. [56] substantiated these findings by showing that the number of MBN pulses varies inversely with grain size in annealed nickel. Large number of pulses means smaller grains, therefore more pinning sites, and possibly more defects.

Gatelier-Rothea et al. [57] reported a decrease in MBN signal when the grain size in iron samples increased. That grain size is inversely proportional to the detected MBN signal was also noticed by Tiito et al. [58] who investigated the magnetic behavior of steel specimens of varying grain size. However, precipitates and segregation of phosphorus at grain boundaries can act as additional pinning sites for domain walls, increasing the number of MBN pulses even in large grained specimens, as observed experimentally in decarburized steel [56].

2.4.2 Compositional and Phase Influences

Compositional variations can bring their own contribution to signals, rendering the MBN analysis more complicated. Plain carbon steels were studied by Kameda et al. [59], who found that variations in MBN signal are obtained because of phase changes such as carbide precipitation or intergranular impurity segregation left behind after heat treatment. Jiles [60] investigated plain carbon steels of the AISI 10xx series, documenting how MBN changes with carbon content due to increased pinning of domain walls by carbide particles. When the latter built networks of lamella carbide, they provided stronger pinning, as opposed to the spheroidized carbides that impeded less the movement of domain walls. Heat treatment of carbon steel AISI 4130 produces pearlite, bainite, and martensite, each with its distinct MBN signature [60].

Blaow et al. [61] compared MBN signal profiles in cementite, pearlite, and martensite specimens with and without compressive strain. When the specimens were undeformed, MBN signals were more pronounced in the magnetically soft spheroidized cementite, however reduced in pearlite or the magnetically hard martensite specimen tempered at 400°C. The latter also displayed multiple MBN peaks with strain, whereas pearlite and a martensite specimen tempered at 180°C showed only a single peak. However, that peak increased systematically with strain. It should be noted that the MBN parameters employed in these studies are quite different from the ones described above in earlier sections. Instead, magnetization curves were recorded, and it is on these curves that one or more peaks were noticed.

The most significant change with strain in MBN signal profiles in the Blaow et al. [61] study was observed in the spheroidized cementite specimens. A single peak in the undeformed state was followed by peak broadening with three overlapping peaks. It is assumed that cementite lamellas provide strong directional pinning to domain wall motion, as previous reports indicate [62,63]. All MBN signal changes were reversible when loads were removed while still in the elastic stage of deformation [61].

2.4.3 MBN Behavior in Different Materials

Three peak MBN profiles were also observed in the magnetization curves of unstrained mild steel [64] and were attributed to spike (or residual) domains. These are usually nucleated and annihilated in the knee region of the hysteresis loop, but some are retained in the specimen upon saturation. Because some spike domains are retained, reversal of the field causes abrupt nucleation of new domains when the knee region is reached again [64]. Around saturation, the incomplete annihilation of these spike domains causes some MBN activity, as Ng et al. [54] pointed out when they observed a second peak at higher fields in the magnetization curve of low carbon steel.

Buttle et al. [65] investigated the magnetization of quite different material types and observed a single peak in a cold-worked Ni sample, however noticed a three peak MBN profile in an Fe sample. Interestingly, Thompson et al. [66] and Lo et al. [67] reported double peak MBN signals in ferritic–pearlitic steel with higher volume fraction. A second peak in the MBN signal was also noticed by Kleber et al. [68] who applied a compressive elastic and later plastic stress to mild steel as well as Armco iron specimens. Nevertheless, Armco iron displayed different MBN behavior than mild steel.

In the experiment of Kleber et al. [68] an increase in MBN signal was detected with tensile as well as compressive plastic stress in Armco iron specimens [68]. The magnetic behavior seen in Armco iron was ascribed to dislocation tangles and their interaction with magnetic domain walls [68]. On the other hand, mild steel displayed a decrease in MBN with tensile plastic stress, and almost no change in compression, contrary to earlier reports by other groups [25].

Kleber et al. [68] concluded that residual internal stresses play a different role in influencing MBN behavior in tensile vs. compressive deformation in mild steel. By this argument, they attributed the changes in MBN observed in the mild steel specimens to residual stresses. In contrast, dislocation effects on MBN were assumed to be independent of the sign of plastic strain. Nevertheless, the overall objective of the Kleber et al. study [68] was to separate the effect of dislocations from that of residual internal stresses in the plastic regime, hence two materials with different yield strengths were chosen.

A somewhat related goal was targeted by Moorthy et al. [69] in their work on the effects of fatigue and overstressing on the MBN response of case-carburized En36 steel. Given the type of steel and its mechanical processing, these specimens had a sharp change in microhardness with depth level, so that a crack initiated at the surface propagated very fast into the material. Furthermore, dislocations had a small chance to form prior to crack formation, as smaller stress levels did not allow dislocation initiation, while crack formation preceded larger stresses necessary for developing dislocations. Of course, the magnetic behavior was observed to also vary under these circumstances.

In Moorthy et al.'s study [69], MBN peak height increased in the vicinity of the crack, hence this technique was able to detect the crack location. Nevertheless, crack growth represented only a small fraction in the fatigue life of the case-carburized steel, with specimens failing soon after crack initiation. On the other hand, detection of residual stresses proved to be a more useful indicator of impending failure, since the MBN technique was able to better assess the maximum level of bending stress prior to crack initiation. Alerting users before cracks have a chance to form is definitely a more favorable alternative to finding out that the component is about to fail as soon as a crack is detected [69].

The MBN response of nonoriented (3 wt.%) Si–Fe was investigated by V.E. Iordache et al. [70] who subjected the steel specimens to uniaxial tensile stress beyond the macroscopic elastic limit, unloaded the specimens, and performed MBN measurements again during a second reloading. The study allowed a comparison of the different aspects of tensile deformation, from elastic strain

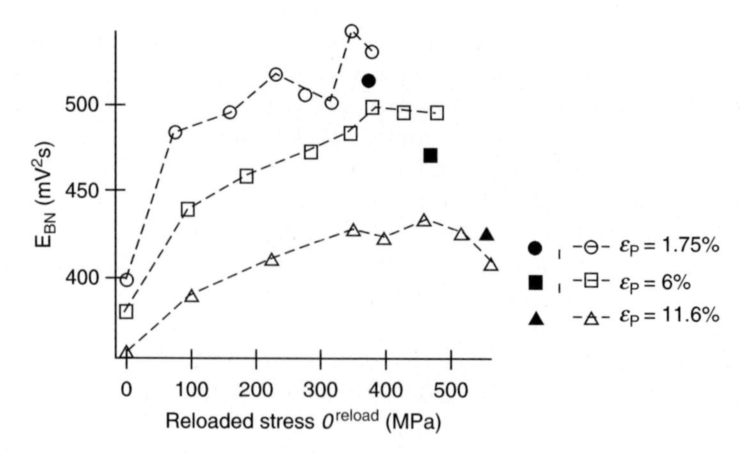

Fig. 2.10. Barkhausen noise energy vs. reloaded stresses (*open marks*) compared with values under initial applied stress (*full marks*) for three nonoriented (3 wt.%) Si–Fe specimens (reprinted from [51] (copyright 2003) with permission from Elsevier)

to microyielding, plastic yielding, and strain hardening. The measurements performed in situ during reloading (Fig. 2.10) revealed some similarities to previous studies of the magnetic behavior of steel, even though the latter were not performed after loading for a second time [20,25]. It should be noted that loading, unloading, and reloading again increases MBN values measured at the same level of stress as during the first loading [71].

2.5 Competitiveness of MBN in Nondestructive Evaluation

2.5.1 Usefulness of MBN for MFL

MFL is an effective inspection method for evaluating corrosion and defect impact on pipelines while they are still in service [72]. It is based on inducing a magnetic flux into the walls of a pipeline, using strong permanent magnets. The flux will "leak out" if metal loss exists, such as when the walls contain defects. A Hall sensor detects the leaking flux that depends on defect geometry, as well as any associated stresses [73]. Given the complexity of the factors influencing the MFL signal, MBN can be of assistance in giving a correct interpretation, provided results from both techniques are compared to establish a common pattern under similar circumstances [73].

Quite often, simulation tools relying on finite element analysis have been employed to model magnetic flux patterns around corrosion defects [74]. Results from MFL studies [73] indicate that flux signals around the defect, in particular dents, are sensitive to elastic residual stresses, but not to plastic deformation. This is in agreement with MBN studies on elastic vs. plastic

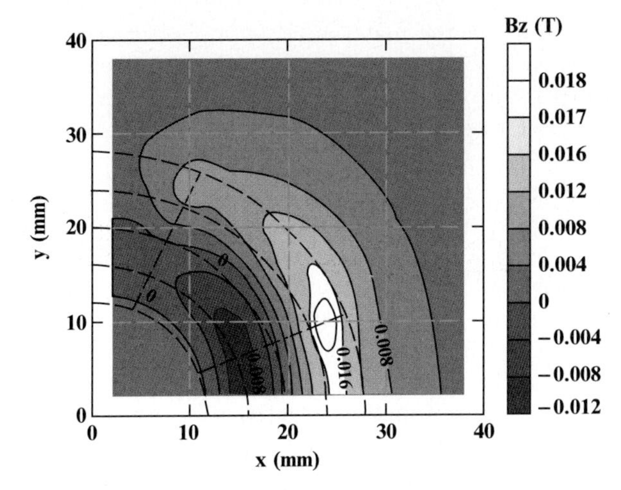

Fig. 2.11. Simulated MFL contour maps obtained 0.5 mm underneath a dented steel plate. The maps show the combined effect of dent geometry and stresses (reprinted from [75] (copyright 2005) with permission from Elsevier)

deformation effects, which also indicate that elastic stress has a more pronounced influence on magnetic behavior than plastic stress [20, 25].

A recent MFL study [73] revealed that the radial signal is most determined by compressive stress configurations perpendicular to the flux path, lowering steel permeability and forcing flux out of the pipe [73]. Figure 2.11 shows the simulated magnetic field contour maps on the bottom side of a steel plate containing a dent, illustrating the effect of both dent geometry and stress configurations on the MFL signal. It seems that geometry is the dominant factor, and not stress distributions around the dent, except for a very pronounced peak in the lower part of the figure. The peak is attributed to stress and correlates well with previous findings of MBN studies [20, 25, 75].

The MFL simulation results [73] were correlated with experimental magnetic flux patterns determined using the MFL technique. Specimens were subjected to a residual stress removal treatment before a dent was applied, to ensure that only stress configurations associated with the dent itself were present. Experimental results matched very closely those obtained by simulation [73]. The knowledge gained through prior MBN studies on stress configurations and magnetic behavior assisted in obtaining a more complete interpretation of MFL signals [73].

2.5.2 Need for Calibration of MBN as NDT

The main advantage of magnetic Barkhausen noise as a nondestructive technique is that it reveals complex changes in materials structure that originate in microstructural properties, as well as stress configurations, together leaving their imprint on the material. How the latter responds depends on multiple factors, each with a specific influence on magnetic behavior. Nevertheless, these factors are difficult to dissociate and treat individually, requiring an experienced user to give the appropriate interpretation to the measured MBN data.

For instance, magnetic domains of neighboring grains of the same crystallographic orientation could become simultaneously active. In this case, an enhanced MBN signal is likely to be obtained, however this large signal may not be entirely because of crystallographic texture. Furthermore, reorientation of grains that can occur with plastic deformation is likely to cause redistribution of internal stresses, potentially destroying the effects of other factors such as residual stresses. Fortunately, crystallographic reorientation only happens at high deformation levels, usually associated with mechanical processing [44].

User knowledge and experience are extremely important in MBN data analysis. Nevertheless, because MBN depends on so many influential variables, it is still necessary to resort to comparative studies, and ultimately calibrate measurements. An undeformed specimen or a known microstructural state needs to act as a standard for a particular type of alloy. In spite of commonalities, one should not directly compare MBN results for different types of plain carbon steels. Their prior magnetic history, processing treatment and ultimately type of steel will determine their MBN response.

References

1. R.M. Bozorth, *Ferromagnetism* (IEEE Press, New York, 1951)
2. W.A. Theiner, H.H. Williams, Determination of Microstructural Parameters by Magnetic and Ultrasonic Quantitative NDE, in *Nondestructive Methods Material Property Determination* (Plenum Press, New York, 1984), p. 249
3. S. Chikazumi, *Physics of Magnetism* (Wiley, New York, 1964)
4. V.N. Shah, P.E. MacDonald, *Residual Life Assessment of Major Light Water Reactor Components*, vol. **1** (Idaho National Engineering Laboratory, Seattle, 1987), p. 144
5. S. Nath, B. Wincheski, J.P. Fulton, M. Namkung, IEEE Trans. Magn. **30**(6), 4644 (1994)
6. R.A. Wincheski, M. Namkung, Aerospace Am. **36**(3), 27 (1998)
7. B. Wincheski, J.P. Fulton, S. Nath, M. Namkung, J.W. Simpson, Mat. Eval. **52**(1), 22 (1994)
8. V. Babbar, B. Shiari, L. Clapham, IEEE Trans. Mag. **40**(1), 43 (2004)
9. R. Becker, W. Döring, *Ferromagnetismus* (Springer Verlag, Berlin, 1939)
10. H. Barkhausen, Physik. Z. **20**, 401 (1919)
11. A.E. Lord, in *Acoustic Emission*, ed. by W.P. Mason, R.N. Thurston. Physics Acoustics, vol. 9 (Academic Press, New York, 1975)
12. M. Namkung, S.G. Allison, J.S. Heyman, IEEE Trans. Ultrasonics Ferroelectrics Freq. Control **33**(1), 108 (1986)
13. K. Tiito, in *Nondestructive Evaluation: Application to Materials Processing*, ed. by O. Buck, S.M. Wolf (ASM, Materials Park, Ohio, 1984), p. 161

14. X. Kleber, A. Vincent, NDT&E Int. **37**, 439 (2004)
15. B.D. Cullity, *Introduction to Magnetic Materials*, 2nd edn. (Addison-Wesley, New York, 1972)
16. M. Blaow, J.T. Evans, B. Shaw, Mater. Sci. Eng. **A386**, 74 (2004)
17. D.C. Jiles, *Introduction to Magnetism and Magnetic Materials* (Chapman and Hall, New York, 1991)
18. J.C. McClure, Jr., K. Schröder, CRC Crit. Rev. Solid State Sci. **6**, 45 (1976)
19. W. Heisenberg, Z. Physik **49**, 619 (1928)
20. C-G. Stefanita, Ph.D. Thesis, Department of Physics, Queen's University, Kingston, Ontario, Canada, 1999
21. D. Utrata, M. Namkung, Rev. Progr. Quant. Nondestr. Eval. **2**, 1585 (1987)
22. R.S. Tebble, Proc. Phys. Soc. Lond. B **68**, 1017 (1955)
23. J-K. Yi, *Nondestructive Evaluation of Degraded Structural Materials by Micromagnetic Technique*, Ph.D. Thesis, Department of Nuclear Engineering, Korea Advanced Institute of Science and Technology, Taejon, Korea (1993)
24. H. Kwun, G.I. Burkhardt, *Electromagnetic Techniques for Residual Stress Measurement*, 9th edn. Metals Handbook, vol. 17 (ASM International, Materials Park, 1989), p. 159
25. C-G. Stefanita, D.L. Atherton, L. Clapham, Acta Mater. **48**, 3545 (2000)
26. C. Kittel, J.K. Galt, Solid State Phys. **3**, 437 (1956)
27. T.W. Krause, L. Clapham, A. Pattantyus, D.L. Atherton, J. Appl. Phys. **79**(8), 4242 (1996)
28. T.W. Krause, L. Clapham, D.L. Atherton, J. Appl. Phys. **75**(12), 7983 (1994)
29. C. Jagadish, L. Clapham, D.L. Atherton, IEEE Trans. Magn. **25**(5), 3452 (1989)
30. C. Jagadish, L. Clapham, D.L. Atherton, NDT Int. **22**(5), 297 (1989)
31. C. Jagadish, L. Clapham, D.L. Atherton, J. Phys. D: Appl. Phys. **23**, 443 (1990)
32. C. Jagadish, L. Clapham, D.L. Atherton, IEEE Trans. Magn. **26**(1), 262 (1990)
33. T.W. Krause, K. Mandal, C. Hauge, P. Weyman, B. Sijgers, D.L. Atherton, J. Magn. Magn. Mater. **169**, 207 (1997)
34. H. Kwun, G.L. Burkhardt, NDT Int. **20**, 167 (1987)
35. C.-G. Stefanita, L. Clapham, J.-K. Yi, D.L. Atherton, J. Mater. Sci. **36**, 2795 (2001)
36. R.L. Pasley, Mater. Eval. **28**, 157 (1970)
37. L.E. Murr, Met. Trans. **6A**, 427 (1975)
38. K. Tangri, T. Malis, Surface Sci. **31**, 101 (1972)
39. R.M. Douthwaite, T. Evans, Acta Met. **21**, 525 (1973)
40. V. Moorthy, S. Vaidyanathan, T. Jayakumar, B. Raj, B.P. Kashyap, Acta Mater. **47**, 1869 (1999)
41. C.-G. Stefanita, L. Clapham, D.L. Atherton, J. Mater. Sci. **35**, 2675 (2000)
42. A.J. Birkett, W.D. Corner, B.K. Tanner, S.M. Thompson, J. Phys. D: Appl. Phys. **22**, 1240 (1989)
43. W. Six, J.I. Snoek, W.G. Burgers, De Ingenier **49E**, 195 (1934)
44. G.E. Dieter, *Mechanical Metallurgy* (McGraw Hill, New York, 1961)
45. S. Chikazumi, K. Suzuki, H. Iwata, J. Phys. Soc. Jpn. **15**(2), 250 (1960)
46. N. Tamagawa, Y. Nakagawa, S. Chikazumi, J. Phys. Soc. Jpn. **17**(8), 1256 (1962)
47. J.P. Hirth, Met. Trans. **3**, 3047 (1972)
48. M.A. Meyers, K.K. Chawla, *Mechanical Metallurgy: Principles and Applications* (Prentice Hall, New Jersey, 1984)
49. M.A. Meyers, E. Ashworth, Phil. Mag. A **46**(5), 737 (1982)

50. H. Mughrabi, Acta Metall. Mater. **31**, 1367 (1983)
51. V.E. Iordache, E. Hug, N. Buiron, Mater. Sci. Eng. **A359**, 62 (2003)
52. K. Kashiwaya, Jpn. J. Appl. Phys. **31**, 237 (1992)
53. G. Bertotti, F. Fiorillo, A. Montorsi, J. Appl. Phys. **67**(9), 5574 (1990)
54. D.H.L. Ng, K.S. Cho, M.L. Wong, S.L.I. Chan, X-Y. Ma, C.C.H. Lo, Mat. Sci. Eng. **A358**, 186 (2003)
55. H. Sakamoto, M. Okada, M. Homma, IEEE Trans. Magn. **23**, 2236 (1987)
56. R. Ranjan, D.C. Jiles, O. Buck, R.B. Thompson, J. Appl. Phys. **61**(8), 3199 (1987)
57. C. Gatelier-Rothea, J. Chicois, R. Fougeres, P. Fleischmann, Acta Mater. **46**, 4873 (1998)
58. S. Tiito, M. Otala, S. Säynäjäkangas, NDT Int. **9**, 117 (1976)
59. J. Kameda, R. Ranjan, Acta Metall. **35**, 1515 (1987)
60. D.C. Jiles, J. Phys. D **21**, 1186 (1988)
61. M. Blaow, J.T. Evans, B.A. Shaw, Acta Mater. **53**, 279 (2005)
62. M.G. Hetherington, J.P. Jakubovics, J. Szpunar, B.K. Tanner, Philos. Mag. B **56**, 561 (1987)
63. L.J. Dijkstra, C. Wert, Phys. Rev. **79**, 979 (1950)
64. D.G. Hwang, H.C. Kim, J. Phys. D: Appl. Phys. **21**, 1807 (1988)
65. D.J. Buttle, C.B. Scruby, J.P. Jakubovics, C.A.D. Briggs, Philos. Mag. A **55**, 717 (1986)
66. S.M. Thompson, B.K. Tanner, J. Magn. Magn. Mater. **123**, 283 (1993)
67. C.C.H. Lo, C.B. Scruby, J. Appl. Phys. **85**, 5193 (1999)
68. X. Kleber, A. Vincent, NDT&E Int. **37**, 439 (2004)
69. V. Moorthy, B.A. Shaw, P. Hopkins, NDT&E Int. **38**, 159 (2005)
70. V.E. Iordache, E. Hug, N. Buiron, Mater. Sci. Eng. **A359**, 62 (2003)
71. M. Lindgren, T. Lepistö, NDT&E Int. **34**, 337 (2001)
72. W. Mao, C. Mandache, L. Clapham, D.L. Atherton, Insight **43**(10), 688 (2001)
73. V. Babbar, L. Clapham, J. Nondestr. Eval. **22**(4), 117 (2003)
74. N. Ida, W. Lord, IEEE Trans. Magn. **19**(5), 2260 (1983)
75. V. Babbar, J. Bryne, L. Clapham, NDT&E Int. **38**, 471 (2005)

3

Combined Phenomena in Novel Materials

Summary. This chapter is divided into two parts: magneto-optical media and magnetoelectric materials. The first part deals with magnetic recording and optical readout, a branch of magnetism that tends to be under-appreciated, especially nowadays when new developments have entered the recording market. Therefore, the objective is to draw attention to the significance of continuous media and its impressive stability. These media are not limited by thermal reversal effects of single-domain particles. However, they encounter their own challenges to high quality recording or accurate optical readout. Even Barkhausen noise can affect the latter in a negative way.

In the second part, materials are presented where magnetic properties are intertwined with electric phenomena, creating a unique environment for imaginative applications. Some substances exhibit naturally a unique combination of properties, while others can be engineered to display effects not normally found in individual constituents. The range of applications for metamaterials is expanding every year, including magnetoelectric sensors and transducers, read/write devices, or even microwave dielectrics. Nevertheless, the ongoing problem of how to improve the design of magnetoelectric materials with a strong magnetization–polarization correlation remains, requiring a better understanding of the mechanisms leading to the generation of these effects.

3.1 The Interest in Magneto-optical Media

Materials with intermediate values for magnetic anisotropies have traditionally been employed as magnetic recording media. For instance, magnetic tapes made use of ferromagnetic solid solutions such as γ-Fe_2O_3–Fe_3O_4 in the form of acicular particles [1]. The uniaxial magnetic anisotropy due to their shape gave them good coercivity [2] which also depended strongly on the ratio of ferrous to ferric ions. Higher coercivity values were obtained when Fe^{2+}/Fe^{3+} was greater than 0.04, however the type of binder in which the particles were dispersed (e.g., paraffin or epoxy resin) appeared to have a contribution as well [3]. The influence on the coercivity was seemingly because of the stress applied to the powder particles by the surrounding material, especially when the temperature varied, as there is a difference in thermal expansion coefficients of the magnetic particles and the binder [4].

3.1.1 Conventional vs. Continuous Media

In conventional magnetic recording, the magnetic grains forming a bit (Fig. 3.1) are separated from each other because of *exchange coupling* [5] which plays a tremendous role in the properties of the media, and ultimately the recording process. Also, the thermal stability of conventional magnetic recording media depends on the magnetization reversal of single-domain particles under thermal activation [6]. However, the magnetic properties of magneto-optical (MO) media are *continuous*, so that significant differences exist between magneto-optical and conventional tapes.

For MO media, marks are recorded as magnetic domains between which a magnetic wall exists (Fig. 3.2). With an estimated wall width of 8 nm, and a mark width of 170 nm (corresponding to an areal density 100 Gbits in^{-2}), the energy barrier for the wall motion is 320 times larger than the thermal activation energy kT, implying that the thermal stability of the magnetic walls in MO media is higher than that of magnetic perpendicular recording media [7]. Nevertheless, it is possible to obtain discrete entities for MO data storage as demonstrated successfully using two versatile techniques: X-ray lithography and ion beam etching [8]. Using these, regular micrometer-sized square lattices of round dots were patterned on Au/Co/Au (111) thin films with perpendicular easy magnetization axes. Co layers were sandwiched between a 5 nm thick Au capping layer and a 25 nm thick (111) Au buffer layer deposited on a

Fig. 3.1. A sketch of a bit cell formed of single-domain particles separated from each other because of exchange coupling. Bit cells are constituents of conventional recording media

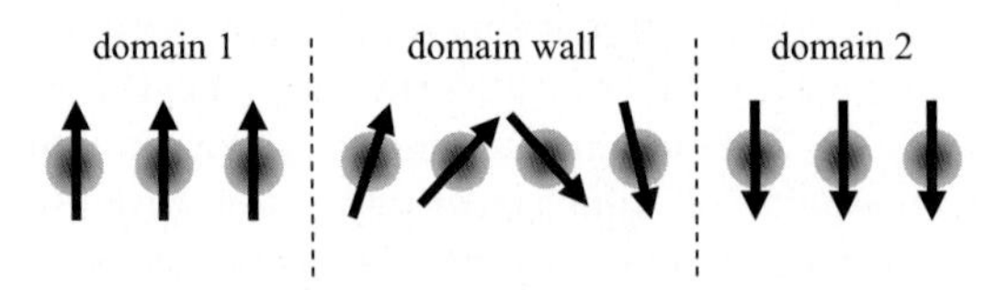

Fig. 3.2. In MO media, marks are recorded as magnetic domains between which a domain wall exists. A domain wall denotes a transition layer which separates two adjacent domains in different directions

thermally oxidized Si substrate. The magnetic dots were obtained through a mask using X-ray lithography, where the unprotected parts of the film were then removed by ion beam etching [9]. The MO Kerr rotation measured on the film showed a strong dependence on the direction of light polarization, and the ratio of film to background reflectance. Single magnetic dots displayed magnetization reversals [10].

3.1.2 The Basis of Magneto-optical Effects

The *Faraday effect* or *Faraday rotation* is a MO phenomenon where the plane of polarization of incident light is rotated proportional to the intensity of the component of the magnetic field in the direction of the beam of light, as the light passes through a material (Fig. 3.3). The effect is observed in *transmission* through a medium, mostly in optically transparent dielectric materials. It was discovered by Michael Faraday in 1845 and is an experimental evidence of the interaction between light and a magnetic field.

The *magneto-optic Kerr effect* is similar to the Faraday effect, however it describes changes in polarization of light that is *reflected* from a magnetized material. This effect takes its name after John Kerr who reported it in 1875 [11].

3.1.3 Composite Films Used in Magneto-optical Recording

Intermetallic compounds, such as MnBi, were among the first materials used for magnetic recording and optical readout by the Faraday–Kerr effect. MnBi has a hexagonal nickel arsenide crystal structure with a very high magnetic anisotropy constant [12]. The direction of easy magnetization is along the c

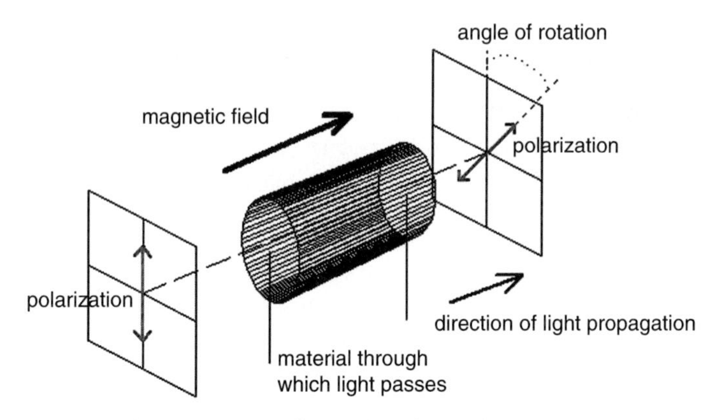

Fig. 3.3. Sketch of the Faraday effect. The plane of polarization of incident light rotates proportionally to the intensity of the magnetic field component aligned with the direction of light propagation. The effect is observed in transmission through a medium

axis. Initially, the Faraday–Kerr effect was used to study magnetic domains in MnBi films [13]. However, once it was noticed that the optical setup can be made such that domains of either polarity appear significantly dark, while those of opposite polarity remained noticeably light; the idea of using this high contrast for magnetic recording and optical readout was born. In fact, the first-reported MO recording was in 1957 [14]. That experiment was based on MnBi magnetic films made by vapor depositing on a glass substrate a layer of Mn, followed by a layer of Bi. The composite film was baked in successive steps at temperatures above 200°C, and sealed off at a pressure of 1.5×10^{-7} mmHg. Over the next 3 days, the film was further baked at temperatures ranging from 225 to 350°C.

Even though the films were not perfectly uniform or continuous, areas of approximately an inch square were crystallographically arranged with the *c* axis normal to the surface, which was also the bulk uniaxial direction of easy magnetization with a spread of only 15°. This detail was particularly important for the optical reading process, as the domain magnetization vectors needed to have components along the direction of propagation of light. Dominant in-plane magnetization vectors would have required oblique illumination. Actually, some previous magnetic domain observation experiments carried out on single crystals of silicon iron and evaporated films of NiFe (80% Ni) used oblique illumination [15,16]. If the films were thinner and more transparent, magnetic domain observation was made with the transmitted light inclined at an angle of 45° to the surface of the film [17]. However in the experiment of Williams et al., the fact that the easy axis of magnetization was normal to the surface allowed optical readout to occur by shining polarized light perpendicularly on the film.

3.1.4 Magnetic Recording and Optical Readout

During optical readout, the light was reflected back by the surface of the film, and the plane of polarization was rotated either clockwise or counterclockwise depending on the magnetic polarity of the reflecting domain areas. By rotating an analyzer, the light from the domains of one polarity was extinguished while the light from those of the opposite polarity was not. This made the former domains appear dark while the latter were light. A Kerr–Faraday rotation of approximately 5° was obtained, giving a good visual contrast. For writing on the film, an electropolished magnetic probe tip with tapered ends was used as shown in Fig. 3.4. One end of the probe wrote dark, whereas the other end wrote light, depending on which direction was magnetized in the local area. If the analyzer is rotated to the other position of maximum contrast, the light and dark areas are reversed, as shown in Fig. 3.4a, b.

3.1.5 Quality of Magnetic Recording

The grey background was due to many fine magnetic domains that appeared either dark or light. They ranged in size from approximately 3,000–6,000 Å,

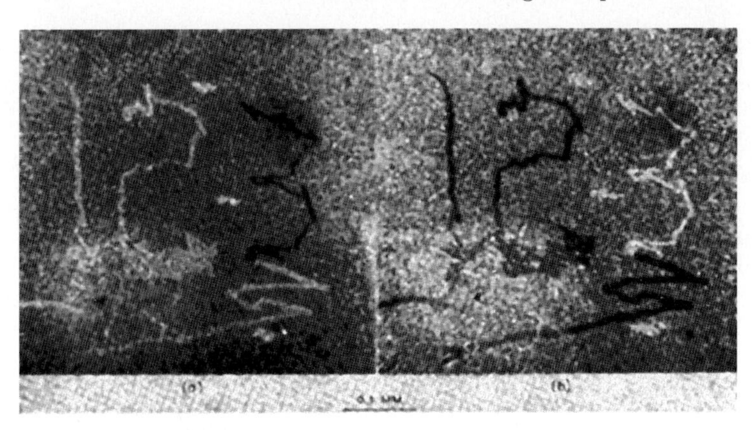

Fig. 3.4. Magnetic writing on a film of MnBi. The numbers 1 and 2 were written with one end of a magnetic probe, and 3 with the other end. Analyzer setting: (**a**) +5° and (**b**) −5° (reprinted from [14] (copyright 1957) with permission from the American Institute of Physics)

and this was also the width of the domain boundaries, revealing a very fine domain structure of adjacent boundaries. It was estimated that 10^6 bits could be recorded on 1 cm^2 of film, an assessment based on the above observations, and the width of the finest line which was 0.001 cm. Nevertheless, the limited capabilities of the read/write equipment would have been an impediment as well. The writing was erased by either saturating or demagnetizing the film perpendicular to the surface. Although the Faraday–Kerr angle of rotation was fairly large, this MO recording method using MnBi films had the disadvantage of large media noise due to the grain boundaries in the crystallite. In spite of its shortcomings, a better understanding of the MO effect in MnBi was still being sought decades later through theoretical studies of the spin–orbit interaction and exchange splitting of the electronic structure [18].

3.1.6 Overcoming Noise Problems

In the 1970s, other media for MO recording were introduced that displayed drastically reduced noise [19]. Among these, amorphous GdCo films [20] were used because they had no grain boundaries. At the same time, it was determined that amorphous rare-earth transition metals [21] are good candidates for erasable and nonvolatile MO disks for which thermomagnetic methods are utilized. However, with the advent of semiconductor lasers as well as improved tracking and focusing servo-techniques, MO disk drives were developed in the 1980s based on amorphous TbFe [22] and GdFeCo as the MO recording medium [23]. By the end of the decade, 130 mm MO disks with a (two side) capacity of 650MB were commercially available. At the beginning of the 1990s, single-sided 90 mm MO disks of 128MB for data recording, and 64 mm rewritable Mini Disks for digital audio recording were put on the market [24].

3.1.7 The MO Sony Disk

A cross-sectional view of a MO disk produced by Sony in the 1990s is schematically shown in Fig. 3.5. Marks are recorded on land areas tracked using grooves, obtained on a polycarbonate substrate using a Ni stamper [25]. Also on the substrate, the four layers are deposited by sputtering, with TbFeCo constituting the magnetic layer of about 20 nm thickness. Two dielectric layers, typically Si_3N_4, are deposited on either side of the magnetic layer to protect it against corrosion [26]. The choice for TbFeCo is based on its good properties such as reasonably high perpendicular magnetic anisotropy, large coercivity ($800\,\mathrm{kA\,m^{-1}}$), low noise, and a moderate Kerr rotation angle. Writing onto the disk is obtained by shining a 10 mW light pulse on a 1 μm diameter area on the magnetic layer, raising the temperature within the light spot by heat absorption. When the raised temperature reaches a critical point (170°C for TbFeCo) where the coercivity is equal to an applied magnetic field for writing, a mark is recorded while the magnetization within the light spot is redirected upwards [27].

Readout from the disk is done through the MO effect originating in the magnetic layer, by shining a 1.5 mW light, weak enough to raise the temperature only slightly, much lower than the critical temperature for recording [28]. For linearly polarized light, the polarization of the reflected light rotates in opposite directions for oppositely directed spins. In conventional detection, these Kerr rotation angles are picked up by a beam splitter that divides them into *s*-polarized and *p*-polarized light before entering two detectors. Thus, recorded marks that advance while the disk is rotating are read out by the light beam focused on the media [29]. The limit of this detection method is given by the recording wavelength p of the marks (Fig. 3.6) which needs to be longer than the optical limit $\lambda/(2\mathrm{NA})$ where λ is the readout light wavelength and NA is the numerical aperture of the objective lens.

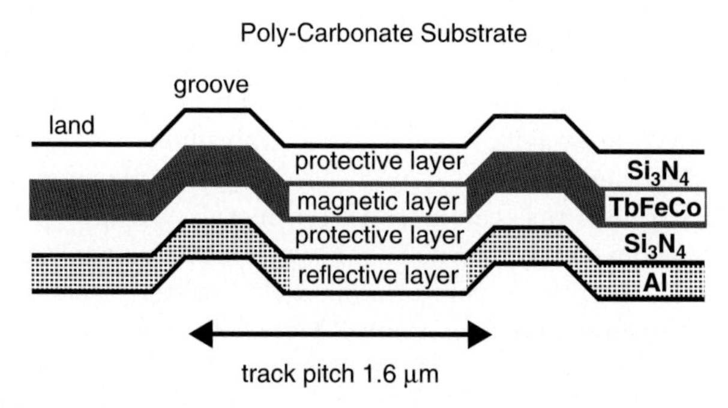

Fig. 3.5. Cross-sectional view of a MO disk from the 1990s (reprinted from [39] (copyright 1994) with permission from the IEEE)

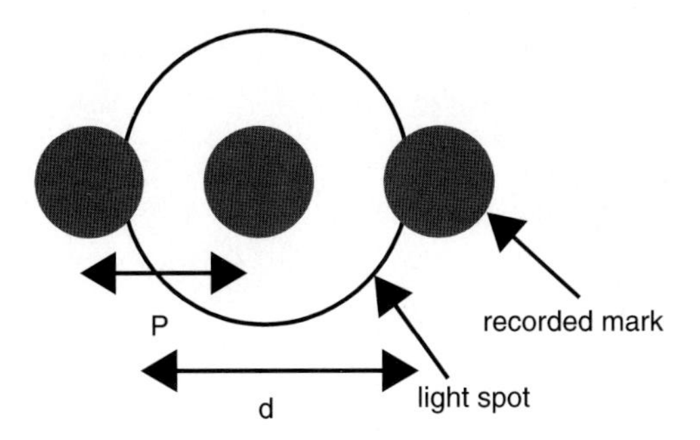

Fig. 3.6. Limit of conventional detection. Recording wavelength p is greater than the optical limit $\lambda/(2\mathrm{NA})$. The first Airy ring has diameter d (reprinted from [39] (copyright 1994) with permission from the IEEE)

3.1.8 Magnetically Induced Super Resolution

To increase the resolution of the optical readout process, the microscope can be equipped with an optical system containing a pinhole smaller than the spot size shining on the object, enabling detection of an object smaller than the limit imposed by conventional detection [30]. However this method, too, has its limitations that can be overcome by having the pinhole produced right in the media. The pinholes are created at the intersection of the readout light spot with a heated area that is produced by a laser ($\sim$10 mW). Thus, a portion of the focused spot is optically masked [31] leaving an effective aperture much smaller than the focused spot.

The signal is optically detected within the effective aperture when a magnetic field higher than the coercivity of the heated region is applied during readout [32]. The magnetization of the heated region aligns with the field, corresponding to erasure [33]. The remaining area apertures the light spot, increasing the resolving power [34] similar to decreasing the light wavelength or increasing the numerical aperture of an objective in conventional detection. This method which expands the readout resolution limit of the MO disk is termed magnetically induced super resolution (MSR) [35]. There are two types of detection by MSR: front aperture detection (FAD) [36] and rear aperture detection (RAD) [37], as shown in Fig. 3.7 [38]. The readout resolution obtained by MSR exceeds the conventional optical limit without the need for a shorter wavelength or a larger numerical aperture of the objective lens [39].

3.1.9 Nondestructive Optical Readout

The increased resolution obtained in the MSR method described above is destructive, as it results in erasing the recorded mark after readout [40]. A non-destructive alternative also developed by Sony was to use an exchange-coupled magnetic triple layer film with a readout, switching and recording layer [41].

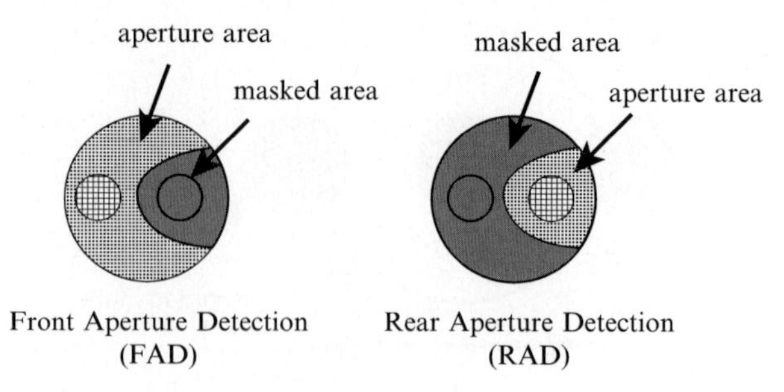

Fig. 3.7. Two types of detection in the MSR method (reprinted from [38] (copyright 1992) with permission from the IEEE)

The readout layer has a low coercivity [42], as opposed to the high coercivity recording layer [43]. When the magnetic layers cool down after the recorded mark in the readout layer is erased by the applied field in the heated area, the information is still retained in the recording layer [44]. This information is subsequently copied onto the readout layer due to the exchange coupling between the readout and recording layers [45], enabling recorded marks to be read out several times [46]. Marks with a recording wavelength of 0.76 mm are resolved for a heating power of 2.5 mW, which is called a resolving power [39].

3.1.10 Double and Multilayer MO Disks

A comparative MO recording study also performed in the mid 1990s at Sony involving Pt/Co multilayer [47] and GdFeCo/TbFeCo double-layer [48] disks using a high power green laser revealed better signal to noise ratios and areal densities three times higher than other MO disks available at the time [49]. The shorter wavelength of the green laser (532 nm), as opposed to the previously used 780 nm laser [50], offered promise of increased densities in these MO disks due to higher coercivities and Kerr rotation angles obtainable in these layers. In the past, TbFe with its large coercivity had proved to be a good choice for the writing layer [51], while GdFe displayed a high Kerr rotation angle very suitable for a readout layer [52]. These exchange-coupled magnetic double layers showed improved carrier levels [53], encouraging further exploration and subsequent development of MO disks with slightly different element combinations: GdFeCo and TbFeCo.

The new films had a combined coercivity higher than 10 kOe. The layers were successively deposited by sputtering on photo-polymerized glass substrates [54] while a specially prepared stamper left sharp grooves with a track pitch of 0.9 μm on the substrate. Scanning tunneling microscopy investigations revealed improved surface smoothness and sharper boundaries between lands and grooves leading to reduced disk noise. The thickness of the layers was optimized for the green laser wavelength. Simultaneous investigations were also performed on Pt/Co multilayers successively deposited by sputtering on the same kind of substrate. The Pt/Co films displayed a relatively high Curie temperature of 400°C and a lower coercivity (1 kOe). Through layer thickness optimization it was possible to obtain at least as good MO properties in these Pt/Co as in the GdFeCo/TbFeCo films.

3.1.11 Domain Wall Displacement Detection

An optical readout technique based on domain wall displacement detection (DWDD) [55] also uses a magnetic triple layer film embedded in a magneto-optical disk [56] (Fig. 3.8). Its strength lies in the fact that its resolving power is not determined by an optical limit but by the driving force of the wall displacement, the latter being temperature dependent [57]. The magnetic triple layer film is exchange coupled so that information recorded on the memory layer is copied onto the displacement layer for readout [58]. However, the exchange coupling is cut off when the switching layer reaches the critical temperature which is almost its Curie temperature. The magnetic domain wall is displaced to the maximum temperature point when the wall passes through a critical temperature contour, also known as the front line (Fig. 3.8).

When the leading edge of the recorded mark passes the front line, a waveform of the signal rises steeply and falls down rapidly as the trailing edge passes the line. The annealed areas on both sides of the track ensure that no side magnetic walls are present [59], a characteristic that is in sharp contrast to the size dependence of magnetic bubble domains whose walls are circumferential [60]. Thus in DWDD, the gradient of the Bloch wall energy, basically independent of domain size, acts as a driving force and is responsible for wall displacement.

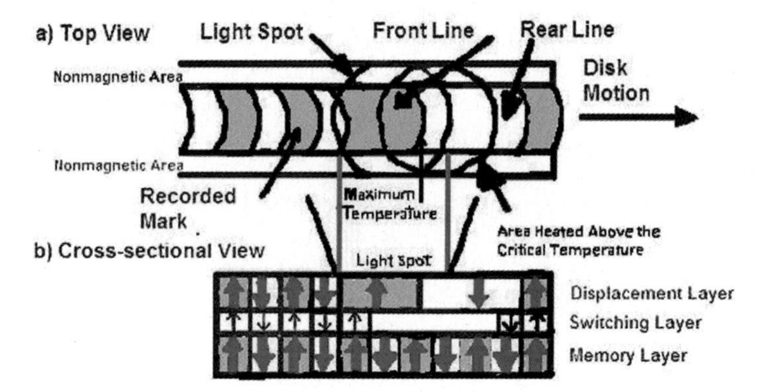

Fig. 3.8. Domain wall displacement detection (DWDD) where light is incident from the substrate (not shown); (**a**) marks recorded by magnetic field modulation (*top view*); (**b**) cross-sectional view of the magnetic triple layer film. The memory layer is TbFeCo, the switching layer is TbFe, and the displacement layer is GdFe (reprinted from [56] (copyright 1999) with permission from the IEEE)

3.1.12 Magnetic Bubble Domains

Magnetic bubble domains are generated in certain materials when the coercivity is lowered under the influence of a laser beam [61]. The coercivity needs to be low compared to the bias field which is necessary for the maintenance of the magnetic bubbles that constitute the written bits [62]. However, when no laser light is shining the coercivity needs to be higher than the bias field to render the material nonvolatile [63]. Only a few milliwatts of power are necessary for writing on a film with low coercivity [64]. In order for the writing to be stable, the bubbles have to nucleate only at the center of the irradiating beam, in the region where the temperature is higher. Thus, bubble formation should be more favorable at higher temperatures because if bubbles were to form in the cooler area of the beam, they would escape the center and writing would become unstable. Materials that satisfy these requirements are Bi containing liquid phase epitaxial garnet films that have a high Faraday rotation angle so that a written bit or magnetic bubble can be read out with a high contrast ratio [65]. These days, Bi substituted garnets are still being studied in detail for technological applications of the MO effect, in particular in optical storage devices [66]. Enhanced Faraday rotation angles obtained in these materials are attributed to different effects, such as mixing of the $3d\,\mathrm{Fe}^{3+}$ orbitals with the $6p\,\mathrm{Bi}^{3+}$ orbitals [67], or the large splitting of the excited state induced by the large spin–orbit coupling of the Bi^{3+} ions [68].

3.1.13 Generation of a Bubble Bit of Memory

A bubble or a bit of memory is generated at a corner of a square formed on the film by etching or ion implantation. Another bubble called a "cue bubble" is also generated to move the bit bubble to the opposite corner. This is done by having the cue bubble follow the movement of the light beam. Due to their low coercivity (typically less than 1 Oe), magnetic bubbles move along the temperature gradient to the region heated by the light beam. As the light beam approaches the bit bubble so does the cue bubble. The bit bubble is repelled by the interbubble repulsive force and is subsequently stabilized at the opposite corner of the square [60]. The size of the written bit is the size of the magnetic bubble domain, and it depends only on the bias field, regardless of the light beam intensity, pulse width, or beam diameter. The bit is circular in shape, irrespective of the shape of the light beam [69].

3.1.14 Driving Force for Wall Displacement

To get an idea about the driving force for wall displacement in DWDD, consider the velocity of the wall. The linear velocity v_{w} of the magnetic wall can be calculated according to the formula

$$\mathrm{v}_{\mathrm{w}} = \mu_{\mathrm{w}}(H_{\sigma} + H_{\mathrm{M}} + H_{\mathrm{ext}} - H_{\mathrm{cw}}), \tag{3.1}$$

where μ_w is the wall mobility, H_σ is the magnetic field caused by the gradient of Bloch wall energy, H_M is the demagnetizing field from the magnetic triple layers, H_{ext} is the external magnetic field, and H_{cw} is the wall coercivity [56]. The latter has a small value for GdFe in the displacement layer when there is no exchange force. Also, the external magnetic field H_{ext} of about 300 Oe is only applied to cancel a ghost signal observed when a longer mark ($> 0.25\,\mu$m) passes the rear line.

On the basis of the above equation, and using some additional theoretical considerations [70, 71] as well as experimental data [72], the velocity of the wall displacement was estimated to be $17\,\mathrm{m\,s^{-1}}$ which compares reasonably well with the experimental value of $28\,\mathrm{m\,s^{-1}}$ obtained in the experiment of Fig. 3.7. The magnetic wall velocity v_w determines the available data transfer rate because the minimum mark length L that can be read out by DWDD is given by the relationship

$$\frac{L}{v} > \tau = \frac{d}{v_w}, \tag{3.2}$$

where v is the linear velocity of the rotating disk ($\sim 1.5\,\mathrm{m\,s^{-1}}$), and d is the displacement of the wall.

Equation (3.2) means that the wall displacement time τ has to be shorter than the time L/ν for the magnetic wall to reach the terminal, before the next wall reaches the front line. Since ν/L gives the data transfer rate, it indicates that the latter is limited by the linear velocity ν_w of the magnetic wall. In this DWDD experiment, the maximum data transfer rate was calculated to be $100\mathrm{MB\,s^{-1}}$ [56]. At the same time, an estimate of the minimum mark length that can be read out by DWDD was $\nu\tau = 50\,\mathrm{nm}$ [73]. It should also be noted that a large wall displacement time τ results in a large jitter, where the jitter in nanometers is given by the jitter in nanoseconds times the disk velocity. The jitter in the time domain is inversely proportional to the maximum slope of a signal wave. But the slope is determined by the wall velocity which becomes a quality factor for the jitter. However, the main reason for the jitter in DWDD remains the fluctuation of the front line.

3.2 Magnetoelectric Materials

3.2.1 The Magnetoelectric Effect

From Maxwell's equations it becomes apparent that electric and magnetic fields are interrelated [74]. A time-varying magnetic flux produces an electric field, and hence a potential drop in a stationary loop. A moving loop with a time-varying area in a stationary magnetic field generates a voltage also known as a motional electromotive force (Faraday's law). Conversely, a steady electric current generates a magnetic field (Biot–Savart law) [75]. These combined electric and magnetic phenomena have resulted in a variety of applications of major technological importance, from traditional electromagnets to the manipulation of magnetic domains by ways other than magnetic fields [76]. In particular, there is an effect that has witnessed a significant revival in recent years. The *linear magnetoelectric (ME) effect* is the induction of a magnetization by an electric field or the induction of an electric polarization by a magnetic field [77]. In essence, in its most general definition, the magnetoelectric effect is the coupling between magnetic and electric fields in matter [78], leading to consequential changes in permeability or permittivity, quite often simultaneously observed with the main effect. The most important requirement for the ME effect to be observed is the simultaneous presence of long-range ordering of magnetic moments and electric dipoles.

3.2.2 Oxides, Boracites, Phosphates, etc.

The ME effect was first predicted by Curie in 1894 [79] and coined as a term by Debye in 1926 [80]. Nevertheless, it was not confirmed until the 1960s when Russian scientists obtained experimental confirmation of an electric field-induced magnetization [81,82] and a magnetic field-induced polarization [83,84] in antiferromagnetic Cr_2O_3. ME materials that caught researchers' attention early on were Ti_2O_3 [85], $GaFeO_3$ [86], garnet films [87], solid solutions such as $PbFe_{0.5}Nb_{0.5}O_3$ [88], boracites [89], and phosphates [90]. These materials must be ordered and have a ferroelectric/ferrielectric/antiferroelectric structure that can undergo a transition to a paraelectric state. Crystal structure plays a significant role in determining the dielectric properties of paraelectric compounds [91]. Conversely, the ferromagnetic/ferrimagnetic/antiferromagnetic state must change to a paramagnetic state. About 80 single-phase materials display the effect [92], yet it is too weak for developing useful applications [93], notwithstanding early predictions [94].

Even with enhanced ME responses observed in some phosphates such as $LiCoPO_4$ [95] and $TbPO_4$ [96] or yttrium iron garnet films [97], microscopic mechanisms are hindering larger induced magnetizations or polarizations in these materials [98], especially at room temperature where ME coupling is small. Additionally, mathematical considerations reveal inherent weaknesses in the ME response [99–101]. Presently, no single-phase material is known that has a high coupling between magnetization and polarization [102], in spite of ongoing research in this area and some progress. For instance, it was noticed that Ti substitution in $BiFeO_3$ renders the compound magnetoelectric, while also refining microstructure and reducing grain size. This effect is attributed to an increase in volume fraction of grain boundaries, leading to an enhanced interfacial polarization [103]. However, improvements in these materials are not significant enough to meet industrial expectations.

3.2.3 Layered Composite Materials

To obtain better results, the solution is to design composite materials formed by two [104] or three [105] constituents such that the ME response would exceed the response observed in single-phase materials [106], thereby advancing the prospects for applications [107]. Tellegen was the first to formulate the idea of such a composite material, although his suggestion was not practically realizable at the time [108]. A composite material is made of layers of magnetostrictive (or piezomagnetic) and piezoelectric constituents [109], in order that a strain produced by a magnetic field applied to the magnetostrictive (or piezomagnetic) material is passed along to the piezoelectric part where it induces a polarization. Most ferromagnetic materials display the magnetostrictive effect. In piezomagnetic materials, the strain induced by a magnetic field is linearly proportional to the field strength, in contrast to magnetostrictive materials where it is proportional to the square of the field strength [110].

Because the layers of the composite material are stress coupled, when the magnetostrictive (or piezomagnetic) part is strained under a magnetic field, an electric field is induced across the piezoelectric part. In magnetostrictive-piezoelectric composites, the ME effect is nonlinear. Although ferrites such as Cr_2O_3 (or a compound containing Cr_2O_3) are not piezomagnetic, a magnetostriction induced in an AC magnetic field gives rise to pseudopiezomagnetic effects, and a ME effect is still observed [111]. The respective thickness and number of layers of the individual phases play an important role in determining the magnitude of the ME response [112], since the compressive stress is higher in thinner piezoelectric layers [113]. Piezocrystals with large piezoelectric voltage coefficients and large elastic compliances give rise to enhanced ME signals. In addition, the volume fraction of the piezoelectric phase needs to be sufficiently high, the piezoelectric coupling coefficient has to be large, and the magnetic phase must have a small saturation magnetization and a high magnetostriction [114].

3.2.4 Product, Sum and Combination Properties

The magnetostrictive and piezoelectric materials in themselves do not always exhibit a ME effect, though a composite made of these materials may [115]. According to the classification given in [116], such a ME composite displays *product properties*, where the ME effect is absent in either of the initial phases. Alternatively, the ME effect in a composite may be due to *sum properties* if it is a result of the weighted contribution of each material, and is proportional to the volume of weight fractions of the magnetostrictive and piezoelectric phases [117]. Similarly, the end compound may display *scaling or combination properties* if the ME response is higher in the composite than in one or more of the constituents [118].

3.2.5 PZT and Magnetostrictive Materials

A commercially available and intensely used piezoelectric material is $PbZr_{1-x}Ti_xO_3$ (PZT) that is often combined with magnetostrictive constituents such as manganites, $LiFe_5O_8$, Permendur and YIG [119]. Nevertheless, the most familiar choice is the ferrite $NiFe_2O_4$ or the doped version

$Ni_{0.8}Zn_{0.2}Fe_2O_4$, as well as compounds where Ni is substituted with Co such as in $CoFe_2O_4$ [120]. Composites with magnetically hard $CoFe_2O_4$ show a weaker ME coupling, in spite of the high magnetostriction observed in $CoFe_2O_4$. This is in contrast to the better coupling obtained in compounds that contain the softer $NiFe_2O_4$ that has a much smaller anisotropy and magnetostriction [121]. Substituting Ni or Co with Zn in different percentages increases the ME response because Zn releases lattice strain, increases permeability, and lowers magnetic anisotropy [122]. An applied magnetic field results in domain wall motion and spin rotation within the magnetic domains; therefore, a high initial permeability or unobstructed domain motion is essential for good ME coupling between the magnetostrictive and piezoelectric phases [123].

3.2.6 Avoiding Ferrites

The low resistivity of ferrites produces leakage currents in the sample, resulting in reduced ME behavior in composite materials. One way to increase the resistivity of the ferrite constituent is to decrease the sintering temperature during the fabrication process [124]. Alternatively, the magnetostrictive behavior observed in $La_{0.7}Sr_{0.3}MnO_3$, $La_{0.7}Ca_{0.3}MnO_3$ [125], or $Tb_{1-x}Dy_xFe_2$ alloys (Terfenol-D) [126] makes them popular substitutes for ferrites as constituents in ME composite materials. Additionally, the ME properties of composites containing a magnetostrictive rare-earth-iron alloy such as Terfenol-D combined with piezoelectric ceramics like polyvinylidine-fluoride (PVDF) have been calculated using the Green's function technique and found likely to exhibit a giant ME effect [127].

Other materials for the piezoelectric part that showed promising results were $Bi_4Ti_3O_{12}, PbMg_{1/3}V_{2/3}O_3, PbMg_{1/3}Nb_{2/3}O_3$–$PbTiO_3$, or $PbZn_{1/3}Nb_{2/3}O_3$–$PbTiO_3$ [128]. Pioneering experiments with these materials [129] revealed that many parameters needed to be tuned to take advantage of the properties of the individual constituents, just as the interactions between them had to be considered to ensure a maximized ME response. By combining ferroelectric piezoelectric materials, such as $BaTiO_3, BiFeO_3$, or $BiFe_{1-x}Ti_xO_3$, and ferromagnetic piezomagnetic $CoFe_2O_4$, researchers were able to obtain a new composite material with an enhanced ME response strongly dependent on the fabrication conditions, usually involving sintering [130]. During this process, solid state reactions take place among the constituent oxides, for instance in the case of $BaCO_3, Co_3O_4, Fe_2O_3$, or TiO_2 when they are heated below their melting point, particles adhere to each other.

3.2.7 Undesired Effects of Sintering

After fabrication, the obtained composite is heated above the ferroelectric transition temperature, and then poled during cooling under an applied electric field. The magnetic poling is done at room temperature under a constant DC magnetic field, usually by keeping the composite samples between the poles of an electromagnet [131]. Sintered composites are less expensive to prepare than unidirectionally solidified in situ materials, in spite of the fact that the ME signal is higher in the latter category. The molar ratio of phases in sintered composites, the grain size of each phase, and the sintering temperature are easily controllable. However, with sintered composites a variety of problems surface, such as the difficulty in passing on the mechanical stress between the magnetostrictive and piezoelectric materials without losses [132] or the reduction in the ME response due to leakage currents in low resistivity magnetostrictive phases. Microcracks due to thermal expansion mismatch between constituents, impurities or undesired phases are also noticed to reduce the ME signal [133].

3.2.8 Variations in Signal Due to Mechanical Coupling

Some variations in ME behavior are observed because of the amount of mechanical coupling between the magnetostrictive and piezoelectric phases [134]. Certain constituents such as $La_{0.7}Sr_{0.3}MnO_3$, $La_{0.7}Ca_{0.3}MnO_3$ or $CoFe_2O_4$ are joined only weakly to PZT [135], while $NiFe_2O_4$ has superior interface coupling to PZT [136]. Additionally, mechanical defects such as pores or microscopic cracks form because of lattice strain, limiting the mechanical bond between the particles of the different components [137]. It was observed that the ME signal increases with higher sintering temperatures for PZT and Ni-ferrites with Fe ions dissolving into the PZT, or the latter diffusing into the ferrite matrix. Doping the Ni-ferrite with Co, Cu, and Mn leads to enhanced values for the magnetostriction, magnetomechanical coupling, and electric resistivity. It should be noted that the dielectric transition temperature in these composites is close to the original dielectric Curie temperature of pure PZT [138].

3.2.9 Laminated Composites

Further improved results were obtained when the magnetostrictive and piezoelectric constituents were assembled from particulate materials [139] or by hot molding, the latter producing interfaces that can no longer be resolved by SEM [140] (Fig. 3.9). Mixtures of PZT/PVDF or Terfenol-D/PVDF are set in molds and subsequently hot pressed [141]. Instead of sintering the constituents, researchers use laminated layers, and indeed enhanced ME responses are observed [142]. An example of a laminated composite is obtained by sandwiching a PZT layer between two Terfenol-D plates and bonding using a conductive epoxy resin, followed by curing the compound at 80°C for 3–4 h under load [143]. The large ME responses given by laminated composites can be attributed to the lack of chemical reactions between independently prepared constituents, unlike in sintered granular composites or unidirectionally solidified composites where the temperature of processing is too high to avoid any chemical reaction of the phases [144]. In spite of the obstacles encountered in experiments with composite materials, the obtained ME response exceeds the largest values observed in single-phase compounds [145].

Fig. 3.9. SEM cross sectional micrograph of a trilayer structure obtained by hot pressing of three laminated layers of mixtures at 180°C and 10 MPa for 30 min. No obvious interfaces can be observed between these three layers (reprinted with permission from [140], copyright 2003 by the American Physical Society, URL: http://link.aps.org/abstract/PRB/v68/e224103; doi:10.1103/PhysRevB.68.224103)

3.2.10 Voltage Coefficient α

ME behavior is usually determined by using a voltage coefficient α, measured in $\mathrm{V\,cm^{-1}Oe^{-1}}$ and expressed as

$$\alpha = \frac{\partial E}{\partial H}, \tag{3.3}$$

where ∂E is the electric field produced by an applied AC magnetic field ∂H. The produced electric field is measured via the voltage ∂V detected across the sample thickness t. One of the first studies on composites done at Philips Laboratories in the Netherlands resulted in a ME composite obtained by unidirectional solidification of a eutectic composition of the quinary system Fe–Co–Ti–Ba–O. Through critical control over the composition, especially when one of the components is oxygen, the eutectic liquid decomposes into alternate layers of perovskite and spinel phases during the unidirectional solidification process. For the researchers at Philips, an excess of TiO_2 (1.5 wt%) in their composites gave a high ME voltage coefficient of $50\,\mathrm{mV\,cm^{-1}\,Oe^{-1}}$ [146]. In subsequent work on $BaTiO_3$–$CoFe_2O_4$, a ME voltage coefficient as high as $130\,\mathrm{mV\,cm^{-1}\,Oe^{-1}}$ was obtained [147]. These values are in contrast to the $20\,\mathrm{mV\,cm^{-1}Oe^{-1}}$ voltage coefficient obtained earlier by the Philips researchers in Cr_2O_3 single crystals. Since then, much higher values were obtained for the ME voltage coefficient.

3.2.11 Obtaining Improved Voltage Coefficients

As noted earlier, the relationship between the applied magnetic field and the voltage induced in the detection circuit is not linear in compounds made of several constituents [148], as it would be for single-phase materials [149]. To obtain linearity in a composite material, the weak AC magnetic field (up to 10 Oe) needs to be superimposed on a large (up to 10 kOe) DC bias field [150], so that the ME effect over a short range around this bias can be approximated as a linear effect, and the voltage coefficient α varies linearly with the DC bias field [151]. For an enhanced response, the modulation frequency of the applied field needs to coincide with the magnetic [152], electrical [153], or mechanical eigenmodes [154] of the system. If the AC magnetic field is tuned to electromechanical resonance, a significant increase in the ME voltage coefficient α is obtained [155]. Observations show that α displays a strong frequency dependence in the microwave range [156, 157].

In addition to frequency considerations, the orientation of the DC magnetic field and AC probe field, as well as the direction of the induced signal with respect to the former two, determines the magnitude of the ME signal [158]. A longitudinal ME response is obtained when both AC and DC applied fields, as well as the induced signal are parallel to each other [159]. Conversely, a transverse ME response is obtained when the induced electric field is perpendicular to the applied DC and AC magnetic fields [160], which need to be parallel to one another for maximum effect [161]. Ultimately, the ME response depends on the relative orientation of the constituents [162]. Enhanced ME responses in laminated plate composites with transverse polarization and longitudinal magnetization with respect to the applied field were obtained by some researchers [163]. Studies were also done for laminate designs with longitudinally magnetized and longitudinally polarized constituents [164]. Their findings depended on the direction of maximum magnetostrictive strain.

Considering that magnetostrictive strain is anisotropic, the ME effect is less pronounced for a direction with reduced magnetostriction. ME coupling is increased when a small increase in the magnetic field results in a significant increase in magnetostriction, since a key requirement for a strong ME coupling is a large magnetostriction. At the same time, the ME coupling between the piezoelectric and magnetostrictive phases vanishes when the magnetostriction attains saturation [165]. Therefore, any dependence of the ME effect on field orientation results from the magnetostriction.

3.2.12 ME and Nanostructures

Lately, magnetoelectric compounds have been obtained by self-assembly on a nanoscale, a technique which shows promising results in terms of phenomena that can be exploited through nanotechnology. In an interesting study, hexagonal $CoFe_2O_4$ nanopillars embedded in a $BaTiO_3$ matrix were grown perpendicular to a single-crystal $SrTiO_3$ substrate [166]. The average spacing between the nanopillars was 20–30 nm, and the whole composite, nanopillars and matrix materials, was grown with a high degree of crystallographic orientation. In this ME compound, the (quasi)unidirectional confinement of the nanopillars led to strains in the magnetostrictive $CoFe_2O_4$ lattice, deformation that was further enhanced due to structural distortions it created in the ferroelectric $BaTiO_3$ matrix. Good connectivity among the constituents was revealed when temperature-dependent magnetization measurements showed significant coupling between the magnetic and electric phases, coupling that dropped at the ferroelectric Curie temperature of 390 K.

In an alternative experiment, ferromagnetic $NiFe_2O_4$ (NFO) nanoparticles were dispersed with their crystal orientations aligned in a ferroelectric $Pb(Zr_{0.52}Ti_{0.48})O_3$ (PZT) matrix [167]. The NFO-PZT nanocomposite was fabricated as a thin film on (001) oriented 0.5% Nb-doped $SrTiO_3$ (Nb:STO) substrates using pulsed laser deposition. The advantage of having nanoparticles rather than nanopillars is that the geometry of the former reduces possible leakage current paths through the ferromagnetic phase. In the end, measured ME parameters, such as remnant polarization, saturation polarization, coercive electric field, and their magnetic counterparts, were in good agreement with reported values for bulk composites [168, 169]. Furthermore, the magnetoelectric voltage coefficient α displayed hysteresis and remanence in its magnetic field dependence, similar to bulk composites [170].

3.2.13 Effects on a Nanoscale

Previous theoretical methods such as Green's function technique [171], finite element method, [172] or the constitutive equations [173] have been developed only for the calculation of ME responses of bulk composites, considering no influence of residual strain. Even fewer theoretical approaches exist for nanostructured ME compounds [174]. Aside from quantum mechanical phenomena revealed on a nanoscale [175], the magnetic, dielectric, magnetostrictive, and piezoelectric properties are also strongly dependent on mechanical boundary conditions, or mechanical constraints caused by lattice misfit between film and substrate. Choosing appropriate ferroelectric and magnetic materials for an epitaxial nanocomposite requires consideration of lattice match and elastic properties, solid solubility, as well as chemical compatibility.

Growing $CoFe_2O_4$ at elevated temperatures surrounded by ferroelectric $BaTiO_3$ leads to self-assembly of the former into columnar nanostructures due to reduced solid solubility of the two phases (Fig. 3.10a, b). As a result of the lattice match between $BaTiO_3$–$CoFe_2O_4$ and the substrate, good epitaxial alignment is obtained along the film's normal (Fig. 3.10c). Furthermore, magnetism and ferroelectricity coexist in the thin film, as proved by the magnetic force microscopy (MFM) images shown in Fig. 3.11, and piezoelectric force microscopy experiments. These scans were repeated after every electrical poling or magnetizing process. Consequently, it was determined that

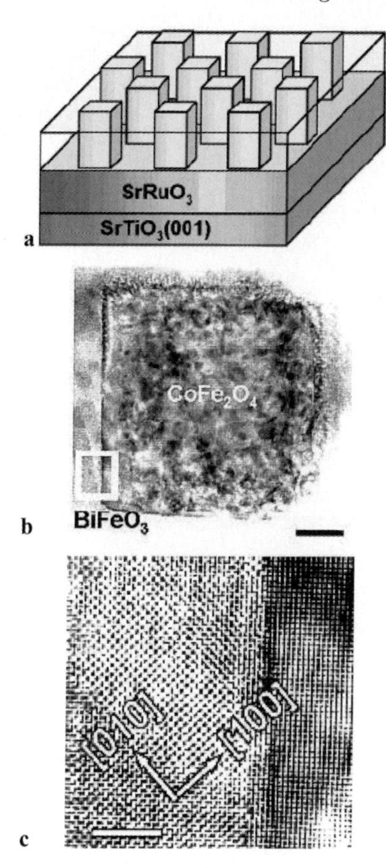

Fig. 3.10. (**a**) Sketch of the self-assembled $BaTiO_3$–$CoFe_2O_4$ heterostructure on a (001) $SrTiO_3$ substrate with a thick $SrRuO_3$ electrode. (**b**) Plan-view TEM image of the $CoFe_2O_4$ nanopillars surrounded by the $BaTiO_3$ matrix. The bar is 30 nm. (**c**) High resolution plan-view TEM image of the $BaTiO_3$–$CoFe_2O_4$ interface taken in the area marked by a *rectangle* in (**b**). The bar is 3 nm (reprinted from [176] (copyright 2005) with permission from the American Chemical Society)

an applied electric field can switch the magnetization between two stable states in the $CoFe_2O_4$ columns. Similar results were obtained for the polarization in the $BaTiO_3$ matrix under an applied magnetic field, proving that there is a strong elastic strain-mediated ME coupling between the two ferroic components of the nanocomposite. Through calculations it was shown that the piezoelectric elastic energy density ($\sim 0.6 \times 10^5\,\mathrm{J\,m^{-3}}$) is of the same order of magnitude as the magnetoelastic energy density ($\sim 0.9 \times 10^5\,\mathrm{J\,m^{-3}}$) that needs to be overcome to enable a temporary change in the magnetic anisotropy [176].

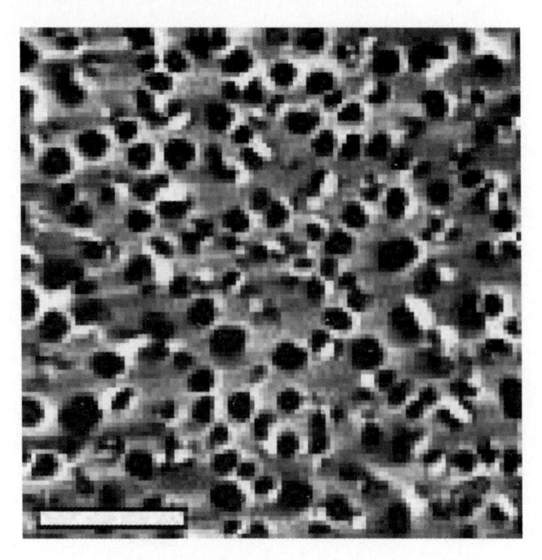

Fig. 3.11. Magnetic force microscopy image after magnetizing the $BaTiO_3$–$CoFe_2O_4$ film perpendicular to the surface in a magnetic field of 20 kOe. The magnetic anisotropy in the columnar $CoFe_2O_4$ is perpendicular to the surface. The bar is 2 μm (reprinted from [176] (copyright 2005) with permission from the American Chemical Society)

3.2.14 Residual Stresses and Strains in Nanostructures

Further experimentation [177] determined that the coupling interaction in a nanostructured multiferroic $BaTiO_3$–$CoFe_2O_4$ film is different from that in bulk composites due to large longitudinal residual stresses and strains created by the $CoFe_2O_4$ nanopillars embedded in the $BaTiO_3$ matrix. These results were backed up by theoretical models based on constitutive equations for piezoelectric and piezomagnetic materials, equations modified on purpose for nanomaterials. The ME parameters calculated from the theoretical models for nanostructured systems agree well with the experimentally measured values, thereby verifying the residual strain induced variations in the ME films. Hence, nanostructured ME compounds can act as a transducer between magnetic and electric field signals, proving that ME materials can be obtained by nanofabrication, while intensifying the anticipation of revolutionary changes in miniaturization, integration, and length-scale reduction in instrumentation and electronics. Newly developed nanomaterials may finally bring the ME effect to practical levels for significant advances in technologies that have undergone promising developments since the 1990s: magnetic–electric sensors and transducers, read/write devices, or even spintronics [178]. Nevertheless, the ongoing problem of how to improve the design of ME materials with a strong magnetization–polarization correlation remains, and it naturally requires a better understanding of the mechanisms leading to the generation of the ME effect.

3.2.15 Multiferroics

Extensive research over the last decade has shown that enhanced signals are obtained in materials with existing or induced long-range magnetic or electric ordering that can in their turn, generate magnetic or electric fields. It was noticed that tuning of ME performance can be achieved in boracites with clamped switching of electrical and magnetic domains, in orthorhombic manganites where a ferroelectric restructuring can be brought about by an applied magnetic field, or in hexagonal manganites with electric field-induced ferromagnetic ordering [179]. These observations led to a new classification for these compounds. They are called *ferroics* and are mainly characterized by the fact that switchable ferromagnetic or ferroelectric domains are formed, and ferromagnetic or ferroelectric ordering can be induced by an applied (or sometimes intrinsic) electric or magnetic field [180]. Furthermore, if these compounds are fabricated such that ferromagnetic and ferroelectric phases are made to coexist, then manipulating the ferromagnetic phase implies that setting or reading of a magnetic state can be obtained by means of a coexisting ferroelectric state, both within the same material [181]. Compounds with two or more phases are known as *multiferroics* [182], since they combine two or more primary ferroic properties.

3.2.16 Using Terfenol-D

Within multiferroic materials, multiple types of domains are present and there is a cross correlation between coexisting forms of ordering. A simple illustration of a three phase ferroic is the Terfenol-D/PZT/PVDF composite obtained by molding and pressing at 10 MPa and 190°C, containing distinct particles of the blended parts as shown in Fig. 3.12. The PZT/PVDF particles constitute the ferroelectric part, while Terfenol-D is the ferromagnetic constituent. Terfenol-D particles display a low resistance, and hence it is required that they are well dispersed in the PZT/PVDF matrix to maintain the electrically insulating property of the composite. Otherwise, the low resistance path created by the Terfenol-D particles will cause the polarization charges to leak via this path. A magnetostrictive change in the Terfenol-D particles is transmitted to the PZT/PVDF matrix inducing polarization charges, giving rise to a ME response [183].

3.2.17 Multiferroic Transformers

The combination of magnetostrictive Terfenol-D and piezoelectric PZT may offer applications in miniature transformers [184]. In such studies, a laminated composite of ring-type geometry containing circumferentially magnetized Terfenol-D and circumferentially polarized PZT was found to display a large ME voltage gain. The design of the transformer is based on the piezoelectric and piezomagnetic equations of state in a radial symmetric-mode vibration, and consists of a four-segment piezoelectric ring layer placed between two

Fig. 3.12. Cross-sectional micrograph of three-phase Terfenol-D/PZT/PVDF composite. The particles above 10 µm are Terfenol-D, the 5 µm particles are PZT, and the irregular phase is polyvinylidine fluoride PVDF (reprinted from [183] (copyright 2003) with permission from the American Institute of Physics)

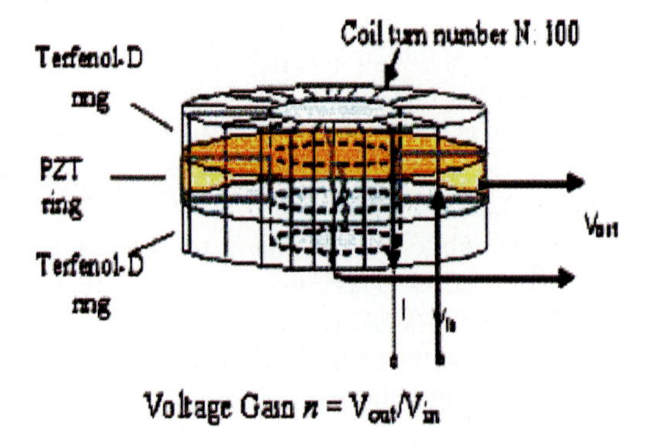

Fig. 3.13. Ring-type ME laminate composite for transformer applications where the voltage gain is defined as $V_{\mathrm{out}}/V_{\mathrm{in}}$ (reprinted from [184] (copyright 2004) with permission from the American Institute of Physics)

magnetostrictive ring layers. Unwanted eddy-currents are minimized or eliminated when the conductive magnetostrictive layers are made very thin. The fields are applied circumferentially, giving rise to strains in the circumferential direction. A sketch of the transformer design is shown in Fig. 3.13.

The AC magnetic excitation field is applied via a current in a toroidal coil that embraces the three ring layers. Hence, the vortex magnetic field is contained circumferentially within the rings, causing the magnetostrictive rings to shrink/expand. As a result, the piezoelectric ring strains radially, producing an output voltage from each of the four segments of the ring. When the frequency of the excitation field is equal to the resonance frequency of the ring and the device functions in resonance drive, the strong ME coupling allows the output voltage to be much higher than the input, exhibiting a large voltage gain. The latter is dependent on the DC magnetic field, and has a value of ~25 at its maximum, for a resonance frequency of 53.3 kHz. Considering these recent advances in ME materials, as exemplified by this transformer that does not require secondary coils with a high-turn ratio, traditional electromagnets may soon be replaced by a solid-state device with a totally different geometry.

3.2.18 Multiferroic Sensors for Vortex Magnetic Fields

Another possible application of ring-shaped laminates of Terfenol-D and PZT is that of sensors for AC rotating or vortex magnetic fields, as described by the same group of researchers mentioned earlier [185]. On the basis of similar construction and design, a vortex magnetic field applied to the Terfenol-D rings gives rise to a strain in the circumferential direction that puts the piezoelectric PZT ring in a radial vibration mode. Detection of the magnetic field occurs with a high sensitivity of 10^{-8} T. Other geometries where, for instance, the magnetic field is parallel, and the electric field is perpendicular to the principal strain direction [186] allow only detection of magnetic fields of constant direction. However, in some cases such as in power-integrated circuits [187] or superconducting films [188] it is the vortex magnetic fields that may require detection.

3.2.19 Enhancing Multiferroicity through Material Design

Multiferroicity can be enhanced or even acquired through proper material design. $PbFe_{1/2}Nb_{1/2}O_{1/3}$ [189] displays a ME effect when transition-metal ions are partly replaced by paramagnetic ions [190]. In a subgroup of hexagonal manganites (e.g. doped perovskite manganites), a magnetic field induces displacement of the Mn^{3+} ions along the hexagonal axis and the subsequent geometric charge ordering generates a ferroelectric polarization, and hence a ME effect [191]. A static magnetic field applied to $Pr_{1-x}Ca_xMnO_3 (0.3 < x < 0.7)$ produces an insulator–metal transition that leads to a transformation from an antiferromagnetic insulating to a ferromagnetic metallic state accompanied by changes in dielectric properties [192]. Rare-earth manganites $RMnO_3$ where R can be Ho, Er, Tm, or Yb can trigger phase transitions and considerably high ME signals. An applied magnetic field induces an antiferromagnetic reorientation of the Mn^{3+} moments and ferromagnetic ordering of the rare-earth sublattices [193]. Sometimes, ME interactions can be strengthened by lowering the porosity and increasing the resistivity in a ferroic compound. Similarly, hot pressing can contribute to controlling key ME parameters that depend on grain size and density, by increasing the initial permeability and strengthening the electromechanical coupling [194].

3.2.20 Identifying Multiferroics

The designation of ferroic depends on nanoscale effects within the material, effects that can ultimately allow the formation of switchable domains, as in ferroelectric or ferromagnetic ordering. The latter are influenced largely by symmetry operations which take place in the crystal when either the polarization or magnetization, or both, rotate. For example, in $Ni_3B_7O_{13}I$ where the polarization and a weak magnetization are coupled, an electric field-induced 180° reversal of the polarization leads to a 90° rotation of the magnetization [195]. This is a symmetry operation of the paramagnetic dielectric phase that leads to a transformation between different domains of the ordered compound. However, it does not further reduce the symmetry of the crystal [196]. On the other hand, a mere rotation of magnetization is not a symmetry operation of the parent phase, and therefore reduces the magnetic energy and increases the free energy of the crystal. Consequently, coupling between polarization and magnetization in a compound conserves the symmetry and makes it possible to obtain a large ME response [197]. It was also demonstrated that in some antiferromagnetic materials spin-rotation domains are generated that supplement the existing spin-reversal domains. In these materials, the piezomagnetic interaction of the strain induced by the ferroelectric domain walls reduces the free energy of the crystal [198], giving rise to ME behavior even if the global symmetry of the crystal forbids the effect. The presence of an external field is usually necessary for bringing about a ME magnetization or polarization. Still, a spontaneous ME effect is possible when an intrinsic polarization causes a weak permanent magnetization [199]. Equally, the magnetic order in some modified perovskite compounds such as $TbMnO_3$ [200] or $DyMnO_3$ [201] can lead to the emergence of polarization giving rise to a ME effect.

Attempts to combine magnetic and electric ordering in the same phase have proved unsuccessful so far, owing to the mutually exclusive nature of atomic-level mechanisms responsible for ferromagnetism or ferroelectricity. On the other hand, the variety of mechanisms responsible for ME behavior in multiphase ferroics clearly indicates that more research is necessary to obtain a comprehensive representation of the sources of the ME effect if better design for these materials is sought.

References

1. P.J. Flanders, IEEE Trans. Magn. **12**, 348 (1976)
2. P.J. Flanders, IEEE Trans. Magn. **10**, 1050 (1974)
3. M. Kaneko, IEEE Trans. Magn. **16**, 1319 (1980)
4. S. Ogawa, C. Handa, S. Ishida, H. Soma, IEEE Trans. Magn. **13**, 1394 (1977)
5. L.R. Bickford Jr., J. Pappis, J.L. Stull, Phys. Rev. **99**, 1210 (1955)
6. H.J. Richter, IEEE Trans. Magn. **35**, 2790 (1999)
7. M. Kaneko, S. Yoshimura, G. Fujita, S. Kobayashi, IEEE Trans. Magn. **36**, 2266 (2000)

8. C. Cesari, J.P. Faure, J. Nihoul, K. Ledang, P. Veillet, D. Renard, J. Magn. Magn. Mater. **78**(2), 296 (1989)
9. F. Rousseaux, A.M. Haghirigosnet, B. Kebabi, Y. Chen, H. Launois, Microelectr. Eng. **17**(1–4), 157 (1992)
10. O. Geoffroy, D. Givord, Y. Otani, B. Pannetier, A.D. Santos, M. Schlenker, Y. Souche, J. Magn. Magn. Mater. **121**(1–3), 516 (1993)
11. A. Yariv, *Optical Electronics*, 3rd ed. (Holt, New York, 1985)
12. C. Guillaud, Thesis, Univ. of Strasbourg, Strasbourg, France, 1943
13. B.W. Roberts, C.P. Bean, Phys. Rev. **96**, 1494 (1954)
14. H.J. Williams, R.C. Sherwood, F.G. Foster, E.M. Kelley, J. Appl. Phys. **28** (10), 1181 (1957)
15. C.A. Fowler Jr., E.M. Fryer, Phys. Rev. **86**, 426 (1952)
16. C.A. Fowler Jr., E.M. Fryer, Phys. Rev. **100**, 746 (1955)
17. C.A. Fowler Jr., E.M. Fryer, Phys. Rev. **104**, 552 (1956)
18. D.K. Misemer, J. Magn. Mater. **72**, 267 (1988)
19. A. Ashkin, J.M. Dziedzic, Appl. Phys. Lett. **21**, 253 (1972)
20. Y. Togami, IEEE Trans. Magn. **18**(6), 1233 (1982)
21. P. Chaudhari, J.J. Cuomo, R.J. Gambino, Appl. Phys. Lett. **22**(7), 337 (1973)
22. Y. Mimura, N. Imamura, Appl. Phys. Lett. **28**(12), 746 (1976)
23. N. Imamura, C. Ota, Jpn. J. Appl. Phys. **19**, L731 (1980)
24. E. Ikeda, T. Tanaka, T. Chiba, H. Yoshimura, J. Magn. Soc. Jpn. **17**(S1), 335 (1993)
25. M. Kaneko, K. Aratani, Y. Mutoh, A. Nakaoki, K. Watanabe, H. Makino, Jpn. J. Appl. Phys. **28**(28), 27 (1989)
26. M. Kaneko, IEEE Trans. Magn. **28**(5), 2494 (1992)
27. A. Okamuro, J. Saito, T. Hosokawa, H. Matsumoto, H. Akasaka, J. Magn. Soc. Jpn. **15**(S1), 447 (1991)
28. Y. Mutoh, T. Shimouma, A. Nakaoki, K. Suzuki, M. Kaneko, J. Magn. Soc. Jpn. **15**(S1), 311 (1991)
29. J. Saito, M. Sato, H. Matsumoto, H. Akasaka, Jpn. J. Appl. Phys. **26**(4), 155 (1987)
30. M. Kaneko, K. Aratani, M. Ohta, Jpn. J. Appl. Phys. **31**(1), 568 (1992)
31. M. Ohta, A. Fukumoto, K. Aratani, M. Kaneko, K. Watanabe, J. Magn. Soc. Jpn. **15**(S1), 319 (1991)
32. A. Fukumoto, S. Kubota, Jpn. J. Appl. Phys. **31**(1), 529 (1992)
33. G. Bouwhuis, J.H.M. Spruit, Appl. Opt. **29**, 3766 (1990)
34. A. Fukumoto, S. Yoshimura, Proc. SPIE **1663**, 216 (1992)
35. S. Yoshimura, A. Fukumoto, M. Kaneko, H. Owa, Jpn. J. Appl. Phys. **31**(1), 576 (1992)
36. F. Maeda, M. Arai, H. Owa, H. Takahashi, M. Kaneko, Proc. SPIE **1499**, 62 (1991)
37. A. Fukumoto, K. Aratani, S. Yoshimura, T. Udagawa, M. Ohta, M. Kaneko, Proc. SPIE **1499**, 216 (1991)
38. S. Yoshimura, I. Nakao, A. Fukumoto, M. Kaneko, H. Owa, IEEE Trans. Consum. Electr. **38**(3), 660 (1992)
39. M. Kaneko, K. Aratani, A. Fukumoto, S. Miyaoka, Proc. IEEE **82** (4), 543 (1994)
40. A. Fukumoto, S. Yoshimura, T. Udagawa, K. Aratani, M. Ohta, M. Kaneko, Proc. SPIE **1499**, 216 (1991)

41. K. Aratani, M. Kaneko, Y. Muto, K. Watanabe, H. Makino, Proc. SPIE **1078**, 258 (1989)
42. A. Nakaoki, S. Tanaka, T. Shimouma, M. Kaneko, Jpn. J. Appl. Phys. **31**(1), 596 (1992)
43. K. Aratani, A. Fukumoto, M. Ohta, M. Kaneko, K. Watanabe, Proc. SPIE **1499**, 209 (1991)
44. A. Nakaoki, Y. Mutoh, S. Tanaka, T. Shimouma, K. Asano, K. Aratani, M. Kaneko, K. Watanabe, Proc. SPIE **1316**, 292 (1990)
45. T. Fukami, Y. Nakaki, T. Tokunaga, M. Taguchi, K. Tsutsumi, J. Appl. Phys. **67**, 4415 (1990)
46. S. Tanaka, T. Shimouma, A. Nakaoki, K. Aratani, M. Kaneko, J. Magn. Soc. Jpn. **15**(S1), 331 (1991)
47. S. Hashimoto, A. Maesaka, Y. Ochiai, J. Appl. Phys. **70**, 5133 (1991)
48. W.B. Zeper, A.P.J. Jongenelis, A.J. Jacobs, H.W. van Kesteren, P.F. Garcia, IEEE Trans. Magn. **28**, 2503 (1912)
49. M. Kaneko, Y. Sabi, I. Ichimura, S. Hashimoto, IEEE Trans. Magn. **29**(6), 3766 (1993)
50. M. Takahashi, J. Nakamura, M. Ojima, K. Tatsuno, Proc. SPIE **1663**, 250 (1992)
51. Y. Mimura, N. Imamura, Appl. Phys. Lett. **28**(12), 746 (1976)
52. S. Tsunashima, H. Tsuji, T. Kobayashi, S. Uchiyama, IEEE Trans. Magn. **17**, 2840 (1981)
53. T. Kobayashi, H. Tsuji, S. Tsunashima, S. Uchiyama, Jpn. J. Appl. Phys. **20**, 2089 (1981)
54. S. Hashimoto, Y. Ochiai, J. Magn. Magn. Mat. **88**, 211 (1990)
55. T. Shiratori, E. Fujii, Y. Miyaoka, Y. Hozumi, J. Magn. Soc. Jpn. **22**(S2), 47 (1998)
56. M. Kaneko, T. Sakamoto, A. Nakaoki, IEEE Trans. Magn. **35**(5), 3112 (1999)
57. T. Shiratori, E. Fujii, Y. Miyaoka, Y. Hozumi, *Digest Int. Symp. Optical Memory* (Tsukuba, Japan, 1998), p. **46**
58. T. Shiratori, E. Fujii, Y. Miyaoka, Y. Hozumi, J. Magn. Soc. Jpn. **23**, 145 (1999)
59. S. Kobayashi, S. Senshu, S. Iwaasa, S. Igarashi, S. Wachi, T. Funahashi, M. Ogawa, Proc. SPIE **1663**, 15 (1992)
60. M. Kaneko, T. Okamoto, H. Tamada, Y. Tamada, IEEE Trans. Magn. **22**(1), 2 (1986)
61. J.P. Krumme, H.J. Schmitt, IEEE Trans. Magn. **11**(5), 1097 (1972)
62. Y.S. Liu, G.S. Almasi, G.E. Keefe, IEEE Trans. Magn. **13**(6), 1744 (1977)
63. H. Callen, R.M. Josephs, J. Appl. Phys. **42**(5), 1977 (1971)
64. A.H. Bobeck, R.F. Fischer, A.J. Perneski, J.P. Remeika, L.G. van Uitert, IEEE Trans. Magn. **5**(3), 544 (1969)
65. W.F. Druyvesteyn, J.Appl. Phys. **46**(3), 1342 (1975)
66. J. Yang, Y. Xu, F. Zhang, M. Guillot, J. Phys: Condens. Matter **18**, 9287 (2006)
67. K. Matsumoto, S. Sasaki, K. Haraga, K. Yamaguchi, T. Fujii, IEEE Trans. Magn. **28**, 2895 (1992)
68. G.F. Dionne, G.A. Allen, J. Appl. Phys. **75**, 6372 (1994)
69. M. Kaneko, T. Okamoto, H. Tamada, T. Yamada, IEEE Trans. Magn. **19**(5), 1763 (1983)

70. O.W. Shih, J. Appl. Phys. **75**, 4382 (1994)
71. E. Schloemann, J. Appl. Phys. **44**, 1837 (1973)
72. M. Tabata, T. Kobayashi, S. Shiomi, M. Masuda, Jpn. J. Appl. Phys. **36**, 7177 (1997)
73. J.C. Slonczewski, J. Appl. Phys. **44**, 1759 (1973)
74. J.D. Jackson, *Classical Electrodynamics*, 2nd ed. (Wiley, New York, 1975)
75. F.T. Ulaby *Fundamentals of Applied Electromagnetics* (Prentice Hall, Upper Saddle River, NJ, 1999)
76. P. Smole, W. Ruile, C. Korden, A. Ludwig, E. Quandt, S. Krassnitzer, P. Pongratz, in *Proc. IEEE International Frequency Control Symposium*, Tampa, 2003, p. 903
77. E. Ascher, in *Magnetoelectric Interaction Phenomena in Crystals*, ed. by A.J. Freeman, H. Schmid (Gordon and Breach, London, 1975)
78. K. Aizu, Phys. Rev. B **2**, 754 (1970)
79. P. Curie, J. Phys. **3**, 393 (1894)
80. P. Debye, Z. Phys. **36**, 300 (1926)
81. D.N. Astrov, Sov. Phys.- JETP **11**, 708 (1960)
82. D.N. Astrov, Sov. Phys.- JETP **13**, 729 (1961)
83. G.T. Rado, V.J. Folen, Phys. Rev. Lett. **7**, 310 (1961)
84. V.J. Folen, G.T. Rado, E.W. Stalder, Phys. Rev. Lett. **6**, 607 (1961)
85. B.I. Al'Shin, D.N. Astrov, Sov. Phys.- JETP **17**, 809 (1963)
86. G.T. Rado, Phys. Rev. Lett. **13**, 335 (1964)
87. T.H. O'Dell, Phil. Mag. **16**, 487 (1967)
88. T. Watanabe, K. Kohn, Phase Trans. **15**, 57 (1989)
89. E. Ascher, H. Rieder, H. Schmid, H. Stössel, J. Appl. Phys. **37**, 1404 (1966)
90. R.P. Santoro, D.J. Segal, R.E. Newnham, J. Phys. Chem. Solids **27**, 1192 (1966)
91. R. Ranjan, A. Agrawal, A. Senyshyn, H. Boysen, J. Phys.: Condens. Matter **18**, L515 (2006)
92. B.B. Krichevtsov, V.V. Pavlov, R.V. Pisarev, A.G. Selitsky, Ferroelectrics **161**, 65 (1993)
93. J.P. Rivera, Ferroelectrics **161**, 165 (1994)
94. V.E. Wood, A.E. Austin, Int. J. Magn. **5**, 303 (1974)
95. J.P. Rivera, Ferroelectrics **161**, 147 (1993)
96. G.T. Rado, J.M. Ferrari, W.G. Maisch, Phys. Rev. B **29**, 4041 (1984)
97. B.B. Krichevtsov, V.V. Pavlov, R.V. Pisarev, JETP Lett. **49**, 535 (1989)
98. C.W. Nan, F.-S. Jin, Phys. Rev. B **48**, 8578 (1993)
99. W.F. Brown Jr., R.M. Hornreich, S. Shtrikman, Phys. Rev. **168**, 574 (1968)
100. R.M. Hornreich, S. Shtrikman, Phys. Rev. **161**, 506 (1967)
101. G.A. Gehring, Ferroelectrics **161**, 275 (1994)
102. S. Dong, J. Li, D. Viehland, Appl. Phys. Lett. **83**, 2265 (2003)
103. M. Kumar, K.L. Yadav, J Phys.: Condens. Matter **18**, L503 (2006)
104. C.W. Nan, M. Li, X. Feng, S. Yu, Appl. Phys. Lett. **78**, 2527 (2001)
105. C.W. Nan, L. Liu, N. Cai, J. Zhai, Y. Ye, Y.H. Lin, L.J. Dong, C.X. Xiong, Appl. Phys. Lett. **81** (20), 3831 (2002)
106. S. Lopatin, I. Lopatina, I. Lisnevskaya, Ferroelectrics **161-162**, 63 (1994)
107. C.W. Nan, Phys. Rev. B **50**, 6082 (1994)
108. B.D.H. Tellegen, Philips Res. Rep. **3**, 81 (1948)
109. C.W. Nan, G.J. Weng, Phys. Rev. B **60**, 6723 (1999)

110. Proc. 3rd Int. Conf. Magnetoelectric Interaction Phen. Cryst. (MEIPIC-3), Ferroelectrics **204**, 356 (1997)
111. K.K. Patankar, S.A. Patil, K.V. Sivakumar, R.P. Mahajan, Y.D. Kolekar, M.B. Kothale, Mat. Chem. Phys. **65**, 97 (2000)
112. G. Srinivasan, E.T. Rasmussen, J. Gallegos, R. Srinivasan, Y.I. Bokhan, V.M. Laletin, Phys. Rev. B **64**, 214408 (2001)
113. A.V. Virkar, J.L. Huang, R.A. Cutler, J. Am. Ceram. Soc. **70**, 164 (1987)
114. S. Shastry, G. Srinivasan, M.I. Bichurin, V.M. Petrov, A.S. Tatarenko, Phys. Rev. B **70**, 064416 (2004)
115. C.W. Nan, M. Li, X. Feng, S. Yu, Appl. Phys. Lett. **78**, 2527 (2001)
116. J. Ryu, S. Priya, K. Uchino, H.E. Kim, J. Electroceram. **8**, 107 (2002)
117. K. Mori, M. Wuttig, Appl. Phys. Lett. **81**, 100 (2002)
118. C.W. Nan, M. Li, J.H. Huang, Phys. Rev. B **63**, 144415 (2001)
119. Proc. 4th Int. Conf. Magnetoelectric Interaction Phen. Cryst. (MEIPIC-4), Ferroelectrics **282**, 1 (2002)
120. M.I. Bichurin, I.A. Kornev, V.M. Petrov, A.S. Tatarenko, Y.V. Kiliba, G. Srinivasan, Phys. Rev. B **64**, 094409 (2001)
121. M. Avellaneda, G. Harshe, J. Intell. Mater. Syst. Struct. **5**, 501 (1994)
122. G. Srinivasan, R. Hayes, M.I. Bichurin, Solid State Comm. **128**, 261 (2003)
123. Landolt-Bornstein, in *Numerical Data and Functional Relationships in Science and Technology, Group III*, Crystal and Solid State Physics, vol. 4(b), Magnetic and Other Properties of Oxides, ed. by K.-H. Hellwege, A.M. Hellwege (Springer, New York, 1970)
124. K.K. Patankar, R.P. Nipankar, V.L. Mathe, R.P. Mahajan, S.A. Patil, Ceramics Int. **27**, 853 (2001)
125. A.P. Ramirez, J. Phys.: Condens. Matter **9**, 8171 (1997)
126. G. Engdahl, Handbook of Giant Magnetostrictive Materials (Academic Press, New York, 2000)
127. C.W. Nan, J. Appl. Phys. **76**, 1155 (1994)
128. M.I. Bichurin, V.M. Petrov, Ferroelectrics **162**, 33 (1994)
129. J van Suchtelen, Philips Res. Rep. **27**, 28 (1972)
130. J. van den Boomgard, R.A.J. Born, J. Mat. Sci. **13**, 1538 (1978)
131. S. Mazumder, G.S. Bhattacharyya, Ceramics Int. **30**, 389 (2004)
132. J. Ryu, A.V. Carazo, K. Uchino, H.E. Kim, Jpn. J. Appl. Phys. **40**, 4948 (2001)
133. S.W. Or, N. Nersessian, G.P. Carman, J. Magn. Magn. Mater. **262**, L181 (2003)
134. N. Nersessian, S.W. Or, G.P. Carman, J. Magn. Magn. Mater. **263**, 101 (2003)
135. M.I. Bichurin, I.A. Kornev, V.M. Petrov, I. Lisnevskaya, Ferroelectrics **204**, 289 (1997)
136. A.M.J.G. van Run, D.R. Terrell, J.H. Scholing, J. Mater. Sci. **9**, 1710 (1974)
137. J. Ryu, S. Priya, A.V. Carazo, K. Uchino, H.E. Kim, J. Am. Ceram. Soc. **84**, 2905 (2001)
138. J. Ryu, A.V. Carazo, K. Uchino, H-E Kim, J. Electroceramics **7**, 17 (2001)
139. G. Srinivasan, C.P. DeVreugd, C.S. Flattery, V.M. Laletsin, N. Paddubnaya, Appl. Phys. Lett. **85**(13), 2550 (2004)
140. N. Cai, J. Zhai, C-W. Nan, Y. Lin, Z. Shi, Phys. Rev. B **68**, 224103 (2003)
141. N. Cai, C.W. Nan, J. Zhai, Y. Lin, Appl. Phys. Lett. **84**, 18 (2004)
142. N. Nersessian, S.W. Or, G.P. Carman, IEEE Trans. Magn. **40**(4), 2646 (2004)
143. S. Dong, J. Li, D. Viehland, J. Appl. Phys. **95**(5), 2625 (2004)
144. S.W. Or, N. Nersessian, G.P. Carman, IEEE Trans. Magn. **40**, 1 (2004)

145. H. Schmid, Int. J. Magn. **4**, 337 (1973)
146. J. van den Boomgard, A.M.J.G. van Run, J. van Suchtelen, Ferroelectrics **14**, 727 (1976)
147. J. van den Boomgard, D.R. Terrell, R.A.J. Born, H.F.J.I. Giller, J. Mat. Sci. **9**, 1705 (1974)
148. K.-H. Chew, L.H. Ong, D.R. Tilley, Appl. Phys. Lett. **77**, 2755 (2000)
149. G. Harshe, J.P. Dougherty, R.E. Newnham, Int. J. Appl. Electromagn. Mater. **4**, 145 (1993)
150. L.H. Ong, J. Osman, D.R. Tilley, Phys. Rev. B **65**, 134108 (2002)
151. J. Zhai, N. Cai, Z. Shi, Y. Lin, C.W. Nan, J. Appl. Phys. **95**(10), 5685 (2004)
152. M.I. Bichurin, V.M. Petrov, O.V. Ryabkov, S.V. Averkin, G. Srinivasan, Phys. Rev. B **72**, 060408(R) (2005)
153. M.I. Bichurin, V.M. Petrov, Y.V. Kiliba, G. Srinivasan, Phys. Rev. B **66**, 134404 (2002)
154. T.R. Tilley, J.F. Scott, Phys. Rev. B **25**, 3251 (1982)
155. S. Dong, J. Cheng, J.F. Li, D. Viehland, Appl. Phys. Lett. **83**, 4812 (2003)
156. K.S. Chang, M. Aronova, O. Famodu, I. Takeuchi, S.E. Lofland, J. Hattrick-Simpers, H. Chang, Appl. Phys. Lett. **79**(26), 4411 (2001)
157. H. Schmid et al. (ed), Ferroelectrics **161-162**, 748 (1994)
158. J.F. Scott, JETP Lett. **49**(4), 233 (1989)
159. M.I. Bichurin, D.A. Filippov, V.M. Petrov, V.M. Laletsin, N. Paddubnaya, G. Srinivasan, Phys. Rev. B **68**, 132408 (2003)
160. G. Srinivasan, E.T. Rasmussen, R. Hayes, Phys. Rev. B **67**, 014418 (2003)
161. Y.N. Venevtsev, V.V. Gagulin, Inorg. Mat. **31**(7), 797 (1995)
162. P. Saintgregoire, Key Eng. Mat. **101**, 237 (1995)
163. M.I. Bichurin, V.M. Petrov, G. Srinivasan, Phys. Rev. B **68**, 054402 (2003)
164. G. Srinivasan, V.M. Laletsin, R. Hayes, N. Paddubnaya, E.T. Rasmussen, D.J. Fekel, Solid State Commun. **124**, 373 (2002)
165. G. Harshe, J.O. Dougherty, R.E. Newnham, Int. J. Appl. Electromagn. Mater. **4**, 145 (1993)
166. H. Zheng, J. Wang, S.E. Lofland, Z. Ma, L. Mohaddes-Ardabili, T. Zhao, L. Salamanca-Riba, S.R. Shinde, S.B. Ogale, F. Bai, D. Viehland, Y. Jia, D.G. Schlom, M. Wuttig, A. Roytburd, R. Ramesh, Science **303**, 661 (2004)
167. H. Ryu, P. Murugavel, J.H. Lee, S.C. Chae, T.W. Noh, Y.S. Oh, H.J. Kim, K.H. Kim, J.H. Jang, M. Kim, C. Bae, J.-G. Park, Appl. Phys. Lett. **89**, 102907 (2006)
168. I. Kanno, H. Kotera, K. Wasa, T. Matsunaga, T. Kamada, R. Takayama, J. Appl. Phys. **92**, 4091 (2003)
169. P. Samarasekara, F.J. Cadieu, Jpn. J. Appl. Phys. **40**(1), 3176 (2001)
170. G. Srinivasan, E.T. Rasmussen, B.J. Levin, R. Hayes, Phys. Rev. B **65**, 134402 (2002)
171. C.W. Nan, L. Liu, Y.H. Lin, Appl. Phys. Lett. **83**, 4266 (2003)
172. G. Liu, C.W. Nan, N. Cai, Y.H. Lin, J. Appl. Phys. **95**, 2660 (2004)
173. G. Harshe, PhD Thesis, The Pennsylvania University (1991)
174. C.W. Nan, G. Liu, Y.H. Lin, H. Chen, Phys. Rev. Lett. **94**, 197203 (2005)
175. P. Bruno, V.K. Dugaev, Phys. Rev. B **72**, 241302(R) (2005)
176. F. Zavaliche, H. Zheng, L. Mohaddes-Ardabili, S.Y. Yang, Q. Zhan, P. Shafer, E. Reilly, R. Chopdekar, Y. Jia, P. Wright, D.G. Schlom, Y. Suzuki, R. Ramesh, NanoLett. **5**(9), 1793 (2005)

177. G. Liu, C.W. Nan, J. Sun, Acta Mat. **54**, 917 (2006)
178. M. Fiebig, J. Phys. D: Appl. Phys. **38**, R123 (2005)
179. H. Schmid, *Magnetoelectric Interaction Phenomena in Crystals* (Kluwer, Dordrecht, 2004)
180. T. Lonkai, D. Hohlwein, J. Ihringer, W. Prandl, Appl. Phys. A **74**, S843 (2002)
181. T.K. Soboleva, Ferroelectrics **162**, 287 (1994)
182. H. Schmid, Ferroelectrics **162**, 317 (1994)
183. C.W. Nan, N. Cai, L. Liu, J. Zhai, Y. Ye, Y.H. Lin, J. Appl. Phys. **94**(9), 5930 (2003)
184. S. Dong, J.F. Li, D. Viehland, Appl. Phys. Lett. **84**(21), 4188 (2004)
185. S. Dong, J.F. Li, D. Viehland, J. Appl. Phys. **96**(6), 3382 (2004)
186. S. Dong, F. Bai, J.F. Li, D. Viehland, IEEE Trans. Ultrason. Ferroelectr. Freq. Control **50**, 10 (2003)
187. G. Busatto, R.L. Capruccia, F. Ianmuzzo, F. Velardi, R. Roncella, Microelectron. Reliab. **43**, 577 (2003)
188. K. Senapati, S. Chakrabati, L.K. Sahoo, R.C. Budhani, Rev. Sci. Instrum. **75**(1), 141 (2004)
189. G.A. Smolenskii, A.I. Agranovskaya, V.A. Isupov, Sov. Phys. – Solid State **1**, 149 (1959)
190. G.A. Smolenskii, I.E. Chupis, Sov. Phys. – Usp. **25**, 475 (1982)
191. C. Ederer, N.A. Spaldin, Nat. Mater. **3**, 849 (2004)
192. Y. Tomioka, A. Asamitsu, H. Kuwahara, Y. Morimoto, Y. Tokura, Phys. Rev. B **53**, R1689 (1996)
193. M. Fiebig, Th. Lottermoser, M.K. Kneip, M. Bayer, J. Appl. Phys. **99**, 08E302 (2006)
194. V.K. Babbar, R.K. Puri, J. Appl. Phys. **79**, 6515 (1996)
195. H. Schmid, Bull. Mater. Sci. **17**, 1411 (1994)
196. V. Janovec, L.A. Shuvalov, in *Magnetoelectric Interaction Phenomena in Crystals*, ed. by A.J. Freeman, H. Schmid (Gordon and Breach, London, 1975)
197. S.M. Skinner, IEEE Trans. Parts, Mater. Packaging **6**, 68 (1970)
198. A.V. Goltsev, R.V. Pisarev, T. Lottermoser, M. Fiebig, Phys. Rev. Lett **90**, 177204 (2003)
199. D.L. Fox, J.F. Scott, J. Phys. C **10**, L329 (1977)
200. T. Kimura, T. Goto, H. Shintani, K. Ishizaka, T. Arima, Y. Tokura, Nature **426**, 55 (2003)
201. T. Goto, T. Kimura, G. Lawes, A.P. Ramirez, Y. Tokura, Phys. Rev. Lett. **92**, 257201 (2004)

4

Magnetoresistance and Spin Valves

Summary. The prospects of uncovering spin dominated transport in structural geometries containing various combinations of magnetic metals/semiconductors/organics have led to a separate and fast-growing research area. While signatures of an effect known as *giant magnetoresistance* (GMR) have been seen in many cases at various temperatures, there is still a fair amount of debate about the origin of the observed results. Additionally, the term *spin valve* has entered mainstream while being applied liberally and interchangeably with GMR, which is strictly speaking not correct. Since magnetic sensors enjoy an increasing popularity in ultra-high density recording applications, it is worth looking at what this is all about. Considering their technological importance and to insure their appropriate employment in a device, magnetic sensors and their underlying structural geometries require an in-depth understanding of the mechanisms triggering resistance changes, as well as the role of accompanying effects such as exchange bias. Not all cases seem to be valid GMR or spin valve structures, nor are all changes in electrical resistance due to spin transport, much less spin filtering. What locks spins at an interface, which electrons pass and which do not are just a few questions not easily answered. For all these reasons and many more, this chapter undertakes to examine closer some of the challenging phenomena that take place in those devices accountable for one of the innovations of today's computer culture: massive information storage.

4.1 Introduction

Interesting effects are unveiled when electron transport is combined with magnetoresistance, in particular when the former occurs through certain coupled multilayer structures composed of magnetic and nonmagnetic metals, semiconductors or even organic materials. Among these effects is the giant magnetoresistance (GMR) for which observations date back to the late 1980s when the first groups [1,2] of scientists to discover GMR did measurements on molecular beam epitaxy (MBE) assembled multilayers. In a varying magnetic field, GMR structures experience changes in electrical resistance when the field overcomes the coupling in their ferromagnetic/nonmagnetic/ferromagnetic configuration.

In a more simplified view, when the applied magnetic field reaches the coercivity of one of the ferromagnetic layers, the magnetization of that layer aligns parallel to the applied field. The magnetoresistance of the system increases or decreases depending on the direction of magnetization in the second ferromagnetic layer. If the magnetizations in the two ferromagnetic layers are parallel to each other the resistance decreases, while the converse is true for antiparallel layers. Upon reaching the coercivity of the second ferromagnetic layer, the magnetization in that layer aligns with the applied field as well, and the magnetoresistance of the system changes again.

4.2 A Simple Way of Quantifying Magnetoresistance

Magnetoresistance (MR) is defined as

$$\mathrm{MR} = \frac{R(0) - R(H)}{R(0)} \times 100\%,$$

where $R(0)$ is the resistance at zero magnetic field and $R(H)$ is the resistance at a magnetic field value H. Depending on the shape of the magnetoresistance curve, the system is said to have *positive magnetoresistance* (i.e., "holds water") or in the converse case, *negative magnetoresistance.* In the 1988 experiment of Baibich et al. [1], the multilayer structures ($\sim$30 layers or more) were formed from (001)Fe/(001)Cr superlattices on (001) GaAs substrates. The same type of ferromagnet existed on both sides of the Cr spacer, and yet GMR was observed. Furthermore, the highest GMR ($\sim$two times resistivity decrease at 4.2 K) was seen in structures with Cr layers only 9 Å thin. Previously, it had been established that (100)Fe/Cr/Fe structures are antiferromagnetically coupled for Cr layer thicknesses of 9 Å, displaying antiparallel magnetizations even in zero field [3]. The fact that GMR is observed for these thin Cr layers is due precisely to this antiferromagnetic coupling between adjacent Fe layers that occurs at Cr layer thicknesses below 30 Å. Nevertheless, there is a limit for how thin these spacer layers can be before oscillatory interlayer exchange coupling sets in and reduces the GMR response [4]. Remarkably, in the Baibich et al. [1] experiment the GMR effect was still observable at room temperature.

4.3 What is Responsible for GMR?

Baibich et al. [1] recognized that spin-dependent transmission of conduction electrons between Fe layers through the thin Cr layers is responsible for GMR, and that the orientation of the magnetization in the Fe layers with respect to the external field plays a role in obtaining GMR. The concept that spin scattering by interface roughness has a noteworthy influence on magnetoresistance with differences in scattering for "spin-up" or "spin-down" electrons was discussed in 1988, acknowledging agreement with earlier experiments [5]. Having a paramagnetic layer with the thickness of a few lattice spacings between two ferromagnetic layers can result in antiferromagnetic behavior of the overall system [6]. Early on [7], IBM introduced a design for GMR sensors in magnetic disk drives consisting of an "artificial" antiferromagnetic structure comprising two ferromagnetic layers antiferromagnetically coupled by means of a Ru layer of only ~3 Å. Because of its thinness, the Ru layer displayed a relatively high interlayer oscillatory exchange energy. Although it ensured a strong exchange coupling, it required a magnetic field of many Tesla to overcome the antiferromagnetic coupling [7].

4.4 Deskstar 16 GP

Multilayer GMR sensors were introduced commercially by IBM in November 1997 as magnetic recording read heads in disk drive products designated [8] Deskstar 16 GP. By using these read heads, extremely small magnetic bits at an areal density of 2.69 Gb in^{-2} could be read. In the initial design of GMR sensors, the current flowed parallel to the layers in the device in a *current-in-plane* (CIP) geometry requiring the sensor to be electrically insulated from the conducting magnetic shields [9]. This would not have been so undesirable if the insulation between the shields would not have occupied space, thereby limiting the recording density of the sensors. When determining the capacity of these early hard disk drives not only the areal density needs to be taken into account, but also the size and number of disks or platters within the drive. Deskstar 16 GP contained 95-mm-diameter disks, each with a storage capacity of more than 3.2 GB, resulting in a total data storage capacity [10] of 16.8 GB.

4.5 "Spin-down" vs. "Spin-up" Scattering: Magnetic Impurities

A magnetic material for which the conductivity depends on spin orientation is likely to contribute to magnetoresistance effects, due to the lower resistivity experienced by one of the spin components. In GMR structures built with these materials, the magnetization vectors in the layers have to be ferromagnetically aligned at some point, so that the spin component with the preferred spin polarization can pass through the layers. Furthermore, impurities need to be present at the interfaces, as they are considered responsible for asymmetric spin scattering [11]. For instance, it was observed that changes in magnetoresistance due to "spin-down" → "spin-down" scattering in Cr impurities in bulk Fe are roughly six times smaller than for "spin-up" → "spin-up" scattering. Because the Fe layers are considerably thicker than the Cr layers, the Cr atoms in the layer of only three lattice constant thickness can be regarded as Cr impurities in bulk Fe. This implies a weak interface scattering and a high transmission for the "spin-down" electrons carrying current with a low resistivity. At zero field, both Cr and Fe atoms (of the layer with antiparallel magnetization) correspond to scattering centers for the mixed "spin-up" and "spin-down" electrons, reducing transmission of electrons through the layers [1].

Similar concepts were introduced also in 1988 by Johnson and Silsbee for spin injection from a ferromagnet [12]. Additionally, theoretical calculations based on the Boltzmann transport equation validated that (Mn, V) impurities achieve comparable scattering effects to Cr impurities in Fe, whereas (Al, Ir) impurities lead to a rapid degradation of GMR [13]. Further experiments in which ultrathin layers of a variety of third elements (V, Mn, Ge, Ir, Al) were deposited between Fe and Cr layers confirmed the importance of spin dependent scattering from impurities at the interfaces [14]. A parameter $\alpha = \rho_{\downarrow}/\rho_{\uparrow} \neq 1$ meant to measure the asymmetry in spin transport through the interface was used for comparison between samples containing different elements. The resistivity ρ corresponded to "spin-down" and "spin-up" electrons, where the arrows were used to distinguish between the two orientations. It was found that a correlation existed between α and the magnitude of the GMR response [15].

4.6 Fabrication of GMR Multilayers: Thin Films and Nanostructures

After 1988, multilayer experiments started to flourish and researchers observed increased GMR values, partly due to other fabrication methods employed for these structures such as magnetron sputtering [16]. In a 1996 experiment [17], a cluster beam deposition process [18] was used to produce high quality Fe/Cu granular films whose GMR was compared to those obtained by a conventional co-evaporation method. The process of cluster beam deposition entails the formation of small metallic clusters by adiabatic expansion of vaporized atoms through a crucible's nozzle in vacuum. An electron beam ionizes the clusters, and a bias voltage is applied between the source and the substrate to propel the clusters onto the substrate while enhancing surface migration of adatoms. To preserve the clusters, an alternative method involves deposition of neutral clusters that are softly landed without ionization or acceleration [17]. In any case, the granular nature of the clusters led to suppressed GMR values, as opposed to the chemically and magnetically homogeneous co-evaporated films. This was because GMR is highly sensitive to nanoscale heterogeneity, and the magnetic state at the interface between the clusters and the host matrix.

Easier magnetization switching, and hence enhanced GMR values have been reported [19] for magnetic layers that are magnetically textured displaying a preferred direction of magnetization. This is the case for layers containing nanowire arrays where the strong shape anisotropy of the latter has proved to be advantageous. It was observed [19] that multilayered films of Fe/$Ni_{80}Fe_{20}$/Co/Cu/Co with a nanowire arrangement display larger GMR values than similar films without nanowires. The researchers attributed this improved response to the shape-induced easy axis of magnetization that allowed straightforward switching of magnetic moments without significant hindrance from the domain structure.

4.7 Spin Valves

The actual variation in magnetoresistance occurs when the field changes the relative orientations of magnetizations in two adjacent magnetic regions. While these magnetizations are parallel to each other pointing in a common direction, electrons with spins oriented in the same direction (for instance, "spin-down") pass easily from one layer to another. Electrons with antiparallel spins (for instance, "spin-up") are strongly scattered, leading to low resistivity for electrons with parallel spins. On the other hand, if adjacent regions have antiparallel magnetizations, both "spin-down" and "spin-up" electrons are strongly scattered, and the resistivity is high for all electrons. This spin-filtering property [20] has led to the term *spin valve.*

A spin valve structure can encompass many layers, however the latter can be divided into three possible categories: (1) a *soft ferromagnet*, appropriately called a *free layer* due to the ease with which the direction of magnetization can be changed; (2) a *spacer layer*, usually a paramagnet; and (3) a *hard ferromagnet* consisting of a *pinned layer* exchange coupled [21] to an *antiferromagnetic pinning layer*, often formed by three layers [22]. Exchange coupling implies that the exchange interaction between magnetic moments is such that the magnetization within the pinned layer cannot change directions easily [23]. The effect was first documented [24] in the 1950s, and has been used in the magnetic recording industry for stabilizing the magnetization within the sensing layer with one of the goals being that of reducing [25] Barkhausen noise.

4.8 The Role of Exchange Bias

Research [26,27] indicates that the exchange bias effect originates from uncompensated interfacial spins locked in the antiferromagnet. Thus, the hysteresis loop of the antiferromagnetic layer is asymmetrical, also said to be shifted horizontally [28]. The spins are therefore "pinned," and do not follow the external field [29]. Specifically, when the bilayer formed by the ferromagnetic and antiferromagnetic layers is cooled below the ordering temperature of the latter, the spins in the antiferromagnetic layer are blocked in antiferromagnetic domains with uncompensated net moments in the direction of the ferromagnetic layer, giving rise to a net magnetic bias field [30].

Studies performed on NiO/Co, IrMn/Co, and PtMn/$Co_{90}Fe_{10}$ using high sensitivity X-ray magnetic circular dichroism spectroscopy in total electron yield detection identified uncompensated Ni or Mn spins located at the antiferromagnetic–ferromagnetic interfaces [31]. The $1/e$ probing depth makes this method responsive to the interfacial surface [32]. By measuring the Ni or Co hysteresis loops, the team [31] was able to identify that roughly only 4% of the ferromagnetic monolayer is tightly pinned to the antiferromagnet, therefore not rotating in an external field. Furthermore, they found that there is a quantitative correlation between the size of the pinned interfacial magnetization and the macroscopic magnetic exchange bias field.

4.9 Ni–Fe Alloys

Generally, Ni–Fe alloys are weak magnetoresistive at room temperature, therefore their study requires low temperatures and fields in excess of 20 kOe. These conditions are disadvantageous in the magnetic recording industry, as it is desirable to have magnetic sensors that detect small fields at room temperature. With the invention of electroplating baths and improved fabrication processes for the heads at IBM, a plated inductive permalloy head [33] was employed in the early days of magnetic recording. This electroplated permalloy was the only selected material for this purpose, as it could meet the strict requirements for both reading and writing processes. With a saturation flux density [34] of 1 T, negligible magnetostriction and corrosion resistance, it was easily plated into films. Its soft magnetic properties made it employable as both read and write element [35].

As the areal density of drives increased, $Ni_{45}Fe_{55}$ started to gain ground. Although its soft magnetic properties were not as good as in permalloy, the increased magnetic moment due to the higher Fe content allowed writing in higher coercivity media. The fact that the magnetic moment for binary NiFe alloys increases monotonically with Fe content up to a composition [34] of 65% Fe, and the better corrosion properties than Co-based high moment alloys led to further exploration of this Ni–Fe alloy. However, trying to optimize the material to improve the writeability process was deteriorating readability, and vice versa. This is because inductive writing usually requires a better magnetic anisotropy, a smaller coercivity and a very low magnetostriction. The problem was solved with the introduction of the magnetoresistive read sensor [36] that was used in conjunction with the inductive write head [37], but with a physical separation between the writer and reader. Once the physical separation between the processes became possible permitting individual optimization of writing and reading, the soft magnetic property requirements of the inductive writer material became less stringent. The head design was improved to handle the positive magnetostriction. After much work leading to serious improvements, $Ni_{45}Fe_{55}$ was accepted as a standard in the thin film head industry after permalloy [38].

4.10 Ternary Alloys

To achieve increased recording densities, higher magnetic moments were needed. Soft magnetic ternary alloys such as CoFeCu with high cobalt content (>75%), low magnetostriction, reduced internal stress and saturation flux densities as high as 2 T were developed by IBM to be used as write heads [39], following permalloy and $Ni_{45}Fe_{55}$. Adding Cu reduced the Barkhausen noise, refined the grain size improving domain configuration, and broadened the range of compositions for which the coercivity is low. Deposition of CoFeCu ternary alloys with uniform composition from a bath containing a reduced concentration of a diffusion-controlled element such as Cu posed challenges during the fabrication process. Adding to the problematic fabrication conditions, the weak corrosion resistance made CoFeCu ternary alloys impracticable in magnetic recording. Furthermore, they did not display a visible advantage over $Ni_{45}Fe_{55}$ in terms of head performance, therefore they were not implemented in magnetic recording.

Preferably, a magnetic recording head should have a high moment, low coercivity, no magnetostriction, large electrical resistance, no internal stress and high corrosion resistance. None of this is achievable in one single material, and compromises need to be made. IBM considered other soft magnetic ternary alloys such as CoNiFe instead of [40] CoFeCu. When electrodeposited galvanostatically from a sulfate/chloride bath similar to the plating bath used for the fabrication of $Ni_{45}Fe_{55}$, the CoNiFe films displayed a magnetic flux density higher than 2 T. More than 250 alloy compositions were obtained by changing the Ni^{2+}, Fe^{2+}, and Co^{2+} concentrations and varying the current density for a given bath composition. However, this CoNiFe plating system was not consistently reproducible in the high moment region of the ternary diagram. More reliable data were obtained for samples with compositions around $Co_{44}Ni_{27}Fe_{29}$ displaying a slightly lower magnetic flux density (~2 T), a coercivity of 1.2 Oe, and a low internal stress (115 MPa). Most samples underwent thermal annealing treatments without damage to their structure [40].

4.11 Ni–Fe Alloys with Higher Fe Content

IBM reexamined higher-iron NiFe binary alloys [41]. Magnetic properties of bulk alloys differ from their electrodeposited counterparts, especially if annealed at high temperatures. Therefore, electrodeposition of NiFe with compositions ranging from $Ni_{35}Fe_{65}$ to $Ni_{15}Fe_{85}$ was performed in sulfate/chloride baths containing boric acid, saccharin, and a surfactant in magnetic fields of ~800 Oe, at temperatures varying between 15 and 30°C according to a patented recipe [42]. Thermal annealing followed at 225°C for 8 h. As a result [43], the films displayed higher values of internal stress and magnetostriction than CoFeNi, nevertheless they were more corrosion resistant and easier to manufacture.

Following further investigations, $Ni_{32}Fe_{68}$ and $Ni_{20}Fe_{80}$ with saturation flux densities of 2 T and 2.2 T, respectively, were adopted [44] by IBM who also investigated the CoFe group of iron alloys. The latter were plated in a galvanostatic bath with a proprietary additive that led to magnetic flux densities of 2.4 T for a 50–70 wt% Fe composition [45]. During the process, bath composition and plating conditions were optimized by investigating the effect of the bath components on the surface pH during plating, as well as current density [46]. These plated alloys with such a high magnetic flux density exhibited a large positive magnetostriction (+45 ppm) and high internal stresses of up to [47] 850 MPa. From permalloy to CoFe alloys, IBM brought significant improvements to magnetic recording heads by developing innovative methods of electrodepositing soft magnetic alloys with magnetic flux densities from 1 to 2.4 T.

4.12 Basic Principles of Storing Information Magnetically

In magnetic recording, information is stored as magnetized regions within thin magnetic layers that have been deposited on glass or aluminum substrates. The flux emanating from the transitions between magnetized regions is detected by a magnetic read head (Fig. 4.1). The strength of the magnetic field at the read head is only a few tenths of Oe, therefore GMR sensors capable of detecting such small fields are important for hard disk drive read head applications [48]. To ensure sensitivity to small magnetic fields, one of the ferromagnets in the GMR structure is magnetically hardened, such that its magnetization is fixed and will not rotate while in the range of fields it is subjected to.

Hardening can be achieved by coupling that ferromagnetic layer to a thin antiferromagnetic layer via exchange biasing, as mentioned earlier. If the layer is pinned, it has a shifted hysteresis loop, due to the exchange bias of the adjacent antiferromagnetic layer. In contrast, the hysteresis loop of the free layer is centered, displaying reduced coercivity. In the end, the strength of the exchange coupling in the reading head needs to be such that the soft ferromagnetic layer can still rotate its magnetization in the presence of a magnetic field, as given for instance by the transition regions in the magnetic tape. As a result, only the free layer magnetization is switched in the low fields created by the bit transitions previously written on the magnetic disk. When the magnetizations of the pinned and free layer change direction from parallel to antiparallel, the magnetoresistance increases due to increased spin-dependent scattering. These changes in magnetoresistance are used to detect transitions between magnetic bits, allowing recovery of stored magnetic data.

The spacer layer thickness of the free layer in spin valves having for instance $Co_{90}Fe_{10}/Cu/Co_{90}Fe_{10}$ pinned/spacer/free layer structures has a

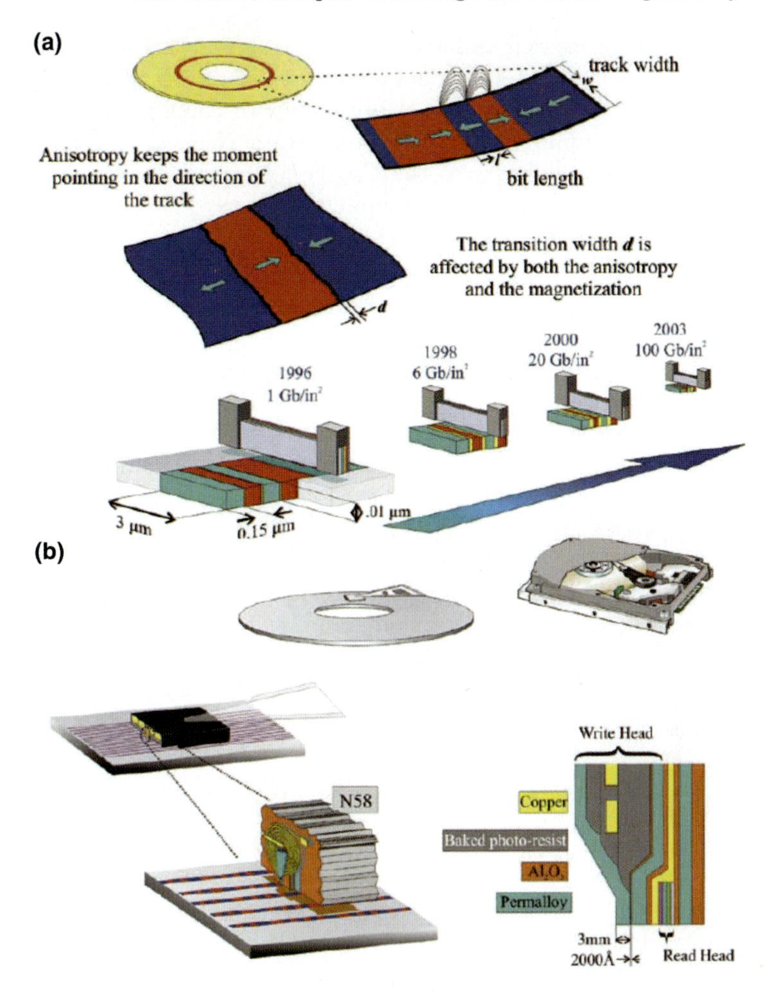

Fig. 4.1. Basic principles of magnetic recording. (**a**) Information is stored as magnetized regions within thin magnetic layers. The flux emanating from the transitions is detected by a magnetic read head while the disk is rotating underneath. The read element scales down in size with the area of the magnetized regions. (**b**) The read sensor is part of a read/write head mounted on a ceramic slider that is flown above a rapidly spinning disk. The hard drive unit typically contains a stack of several head-disk assemblies plus control electronics (reprinted from [48] (copyright 2003) with permission from the IEEE)

noteworthy impact on the interlayer coupling fields [49]. This is because coupling fields depend on the shape of the Fermi surface of the spacer layer, as well as the degree of confinement of the magnetic carriers in the spacer quantum well. In the weak confining system $Co_{90}Fe_{10}/Cu/Co_{90}Fe_{10}$ pinned/spacer/free layer, the minority spin electrons in the spacer layer are confined close to the top of a finite quantum well, influencing considerably the temperature dependence of the oscillatory interlayer coupling constant. Particularly, for PtMn-pinned $Co_{90}Fe_{10}$ layers, Maat [49] et al. found two coupling mechanisms to contribute to the coupling field: the magnetostatic Néel coupling originating from the correlated roughness at the $Co_{90}Fe_{10}$/Cu and Cu/$Co_{90}Fe_{10}$ interfaces, and the oscillatory interlayer coupling.

4.13 Materials for spin valve Sensors

On the basis of various complexities displayed by prototype materials incorporated into magnetic read heads, IBM and later Hitachi GST investigated a variety of materials and material combinations for their suitability as layers part of the their spin-valve sensors. Among these, cobalt ferrite $CoFe_2O_4$ thin films were under consideration as exchange-spring pinning layers [50]. Their magnetic properties are highly dependent on their microstructure which is determined by the fabrication process, in this case, reactive sputtering from a $CoFe_2$ target. Oxide pinning layers are preferred due to their increased corrosion resistance. Although other oxide materials have been considered, such as α-Fe_2O_3 [51], rare earth orthoferrites [52], and $Ni_xCo_{1-x}O$ [53], it is still $CoFe_2O_4$ that has the highest anisotropy constant. Furthermore, orthoferrites are not easily manufactured at room temperature and α-Fe_2O_3 requires thick layers before sufficient pinning is achieved, not to mention they display a low exchange field.

For the spin valve with the $CoFe_2O_4$ pinning layer [54], changes in magnetoresistance of 12.8% and pinning fields of 1,500 Oe have been observed, while the soft magnetic properties of the free layer were maintained. As expected, the spin valve formed of glass/$CoFe_2O_4$/Co/Cu/Co/NiFe/Ru layers displays asymmetric hysteresis loops, as the cobalt ferrite layer is not fully reversed. This is similar to PtMn or NiMn pinning layers, where the coercivity is approximately half of the pinning field value and the $CoFe_2O_4$/Co acts like an exchange spring magnet with saturation fields for $CoFe_2O_4$ exceeding 15 kOe. Nevertheless, the loops become symmetric again for cycles of ± 18 kOe, when the $CoFe_2O_4$ layer experiences full magnetization reversal [54]. If the pinned layer has a too high coercivity it can be hysteretic at moderate fields, and therefore easily demagnetized after several cycles. Additionally, the interactions between the free and pinned layer can render a too high free layer coercivity. Cobalt ferrite spin valves seem to evade these difficulties, as the pinned layer displays no hysteresis, and the coercivity of the pinned layer is similar to the one of antiferromagnetically pinned spin valves.

Further improvements can be made by adding an insulating cobalt oxide (CoO or Co_3O_4) underlayer that pins without shunting current. Aside from its electric insulating properties, this underlayer provides a template for better crystalline growth, maintaining a higher coercivity and thermal stability. Also, free layer properties are preserved, comparable to PtMn based sensors of similar thickness. The insulating layer can be made part of the insulating gap, reducing the thickness of the cobalt ferrite layer to only [55] ∼3 nm. Although the total cobalt oxide/cobalt ferrite thickness is ∼17 nm, it is still under the 40 nm thickness of cobalt ferrite layers that is usually stable. Thus, the spin valve can fit into a 50 nm gap, as required for recording densities of >100 Gbit in^{-2}.

4.14 The Need for Proper Sensor Design

$CoFe_2O_4$ films have a more homogeneous magnetic structure, as compared to hard magnetic materials such as CoPtCr that have a high magnetic anisotropy. The latter apparently leads to fringing fields affecting the free layer, thereby increasing its coercivity. Nevertheless, further studies of $CoPt_{18}$ hard magnet pinning layers (50 Å) on Cr seed layers (>20 Å) fabricated by the IBM team [56], later under Hitachi GST found that free layer softness is not compromised after all if the spacer layer is relatively thick. That, and the reduced critical thickness of the CoPt pinning layer renders the latter a viable alternative to IrMn or PtMn antiferromagnets for small gap sensors in high density magnetic recording. PtMn needs a minimum thickness of 150 Å to become antiferromagnetically ordered upon annealing, while minimum 80 Å are required for IrMn to obtain optimum exchange bias.

The CoPt layers are deposited at room temperature by dc magnetron sputtering in an argon pressure of 2 mTorr using two separate Co and Pt targets. Composition changes are obtained by varying the sputtering power of the Pt source. After using vibrating sample magnetometry and magneto-optical polar Kerr rotation in magnetic field scans up to ±18 kOe, the CoPt magnetization was found to be in-plane, while there was no out-of-plane remanence observed. Furthermore, the in-plane hysteresis loops are alike in any in-plane direction, indicating that the film is magnetically isotropic in plane. For pinning applications in spin valves it is required for the magnetic moment of the CoPt pinning layer to balance the magnetic moment of the reference layer, CoFe. In this study, it was found that the moment of 50 Å of $CoPt_{18}$ equals the moment of ∼38 Å of $CoFe_{10}$. The coercivity and coupling field values of the free soft layer were measured at 5–10 Oe, and the coercivity of the $CoPt_{18}$ layers was found to be ∼1.5 kOe for Pt concentrations 16–22 at.%. The resistivity was measured at only 31 μΩ cm, implying a low parasitic resistance, which is especially important since a pinning layer does not usually contribute to GMR, but rather decreases the overall signal. Its resistance can be similar or greater than the total resistance of the actual spin valve structure. PtMn resistivities between 193 and 227 μΩ cm were recorded, actual values dependent on annealing conditions. Similar resistivity values were obtained [56] for IrMn, 150–162 μΩ cm.

In related investigations by Maat et al. [57] ultrastrong Ir-coupled antiparallel self-pinned layers were used for the stabilization of the thick reference layer. High coupling constants had been previously measured in Ir-coupled layers [58]. The extremely high coupling energies $>3\,ergs\,cm^{-2}$ permit increased layer thicknesses to >100 Å for both pinned and reference layers. The antiparallel pinned structure was balanced in terms of magnetic moments, and high saturation fields were maintained, as required for the stability of the read sensor. The strong coupling was retained even when patterned into nanopillars of 50–200 nm. However, the Ir-coupled spin valves do not display thermal stability, as observed after annealing for 4 h at 255°C.

4.15 Magnetic Tunnel Junctions

Magnetic tunnel junctions (MTJ) have been considered for magnetic recording due to some similarities to spin valve devices. While the sense current is passed not in plane, but perpendicularly through the device, the spacer layer is replaced by a very thin insulating tunnel barrier. Tunneling magnetoresistance responses as high as 50% at room temperature have been reported in devices with alumina tunnel barriers [59]. An anisotropic magnetoresistance arises when the magnetization throughout the device is rotated uniformly so as to change the angle between the direction of current flow and the magnetic moment. The anisotropic magnetoresistance increases significantly when the contact cross section is narrowed to nanoscale, where the point contacts enter the tunneling regime. The magnetic moments of the two contacts rotate together, while remaining parallel to each other [60].

4.16 Anisotropic Magnetoresistive Sensors

Artifacts can arise in MTJs due to magnetostriction effects and magnetostatic forces that can alter the geometry of nanoscale junctions as the magnetic field is varied [61]. These challenges can be overcome by firmly attaching the contacts to a nonmagnetic silicon substrate and performing measurements exclusively at low temperatures, to suppress thermally driven surface diffusion of metal atoms [62]. Nevertheless from a fabrication perspective, a perpendicular current flow is an advantage in magnetic recording, because the sensor can be directly attached to the magnetic shields that can also be used as electrical contacts. Prior to GMR read heads, the magnetic recording industry employed anisotropic magnetoresistive sensors made of two thick, soft magnetic layers shielding a sensing layer in the middle [63]. The spatial resolution of the anisotropic magnetoresistive read head was determined by the thickness of the sensing layer and the separation of the two shields, leading to predictions of maximum areal densities of only $\sim 5\,\mathrm{Gb\,in}^{-2}$ [48].

The anisotropic magnetoresistive heads achieved a low performance of merely a few percent variation in magnetoresistance at room temperature. This was due to the fact that anisotropic magnetoresistance is attributed to bulk scattering within a ferromagnetic metal, with changes in resistance strongly dependent on the angle between an applied external field and the direction of current flow. On the other hand, GMR is predominantly influenced by interface scattering, therefore leading to responses of 65% (at 295 K) and 115% (at 4.2 K) in Co/Cu multilayers in the early [64] 1990s. In some instances, bulk scattering can influence GMR due to the existence of spin-independent imperfections such as lattice distortions, structural disorders or heterogeneity of the material, affecting the mean free path and causing resistance variations under an applied magnetic field even at low temperatures [65]. In particular, ferromagnetic alloys containing permalloy display a random distribution of magnetic atoms causing spin-dependent bulk scattering [66, 67]. However, the latter has a relatively small impact [68] on GMR, as compared to spin-dependent interface scattering.

4.17 Extraordinary Magnetoresistance

In some experimental devices based on a different type of magnetoresistance effect, the so-called *extraordinary magnetoresistance* (EMR), some problems due to the in-plane field sensitivity are avoided, as the geometry of EMR results in sensitivity to magnetic fields perpendicular to the wafer plane. The EMR is modulated by the Lorentz force that acts on the electrons that form a current between two adjacent conduction channels formed by a high mobility semiconductor bar and a low-resistance metallic shunt, both nonmagnetic. EMR devices may be taken into consideration for sensors integrated into planar recording head configurations [69].

4.18 GMR Sensors with CPP Geometry

Meanwhile, GMR sensors with current-perpendicular-to-plane (CPP) geometry are used in the magnetic recording industry. In these, the current travels between the two magnetic shields (Fig. 4.2) perpendicular to the plane of the layers in the spin valve, parallel to the air-bearing surface of the sensor [70]. The spacing between the two shields determines the downtrack resolution of the recorded bits. The overall width of the sensor is responsible for the crosstrack resolution, while the sensor height needs to scale with the track width for good magnetic stability [71]. Increases in GMR can be achieved by partially oxidizing layers within the structure [71, 72]. The total sensor electrical resistance is usually kept low to $\sim 100\,\Omega$, allowing a high sensor bandwidth and a high signal-to-noise ratio. The CPP sensor resistance is characterized by the product: sensor electrical resistance $\times$ track width $\times$ sensor height (in $\Omega\,\mu m^2$), which needs to be kept low. For instance, serial product resistance values (due to the antiferromagnet and excluding underlayers) of only $34\,\Omega\,\mu m^2$ for a PtMn-pinned spin valve, and $13\,\Omega\,\mu m^2$ for an IrMn-pinned spin valve

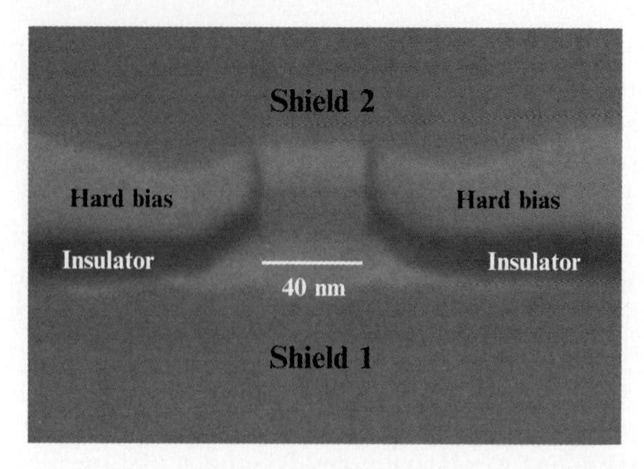

Fig. 4.2. Air-bearing surface view in an SEM micrograph of an all-metal CPP-GMR head with 40 nm reader width (reprinted from [70] (copyright 2006) with permission from the IEEE)

have been obtained [56]. This low product makes CPP GMR sensors more compatible with high density recording heads than MTJ devices that display a too high resistance to be employed in high data rate disk drives [73]. However, parasitic resistance from underlayers, antiferromagnets, cap layers, and other inactive layers reduce the measured GMR.

4.19 Dual Spin Valves

To improve GMR sensor performance, Childress et al. [70, 74] from Hitachi GST investigated a dual spin valve composed of a multilayer stack: underlayer/IrMn[pinned layer/Ru/reference layer]Cu/free layer/Cu/[reference layer/Ru/pinned layer]IrMn/cap (Fig. 4.3)). Thin IrMn (6–7 nm) are preferred to thicker (13–15 nm) PtMn layers, due to the requirement of reduced stack height that influences the head resolution along the recorded track through the total read gap. The Ru layer facilitated the antiferromagnetic coupling between the pinned layer and the reference layer, creating the actual spin valve, and minimizing the magnetostatic coupling to the free layer.

Having a dual spin valve represents a challenge from a magnetic perspective, as the stability of the sensor depends significantly on having the top and bottom pinned layers well pinned. Nevertheless, some difference in pinning strength is expected for the top and bottom pinning layers because of growth conditions, crystallographic and magnetic texture development, and roughness effects. The films were deposited by dc magnetron sputtering onto Si and glass wafers at room temperature [74] or alternatively, on a chemically mechanically polished NiFe shield surface that had been previously electrodeposited onto alumina-TiC substrates [70]. In the latter case, the films were annealed for 5 h at 245°C, then cooled in a 13 kOe magnetic field to set the pinning direction in IrMn. The total stack height of the sensor was kept just

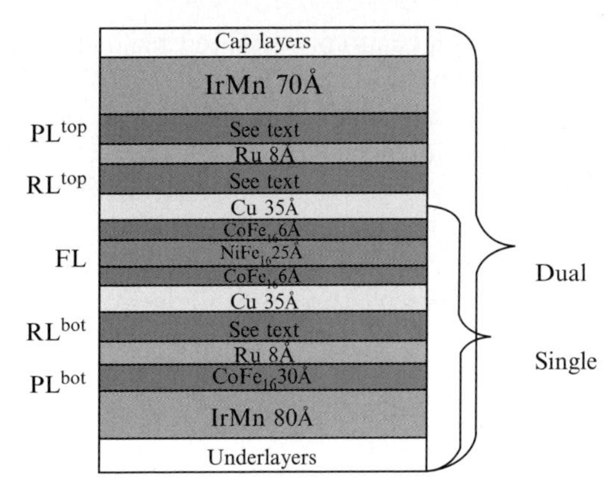

Fig. 4.3. Multilayer stack for single and dual spin valves. The single spin valves are obtained by terminating the structure with a cap after the free layer FL (reprinted from [74] (copyright 2006) with permission from the American Institute of Physics)

under 50 nm [70]. The free layer coercivity of ~9 Oe and the high coupling field of 45 Oe obtained for the dual spin valve design indicated suitability for magnetic recording [74]. Furthermore, this design resulted in stabilized transfer curves in applied fields of ±500 Oe, lower spin torque-induced noise, reduced parasitic resistance and higher GMR values, making them promising candidates for ultra-high density head sensors for recording applications at >300 Gbit in^{-2}.

4.20 Some GMR Multilayer Material Combinations

Ferromagnetic Co layers separated by thin Cu layers were noticed to exhibit GMR of over 110% even at room temperature, compared to only 80% displayed by Fe/Cr structures [75]. Generally, GMR can be observed in a variety of transition metal magnetic multilayers, while the effect fluctuates as the thickness of the spacer layer is varied. Additionally, the shape of the magnetoresistance curve changes with spacer layer thickness [76], as well as due to the fact that the resistance of these materials is strongly temperature dependent. Nevertheless, different thicknesses cause the interlayer exchange coupling to alternate between antiferromagnetic and ferromagnetic, such that the magnetizations of successive ferromagnetic layers are either parallel or antiparallel.

Only antiferromagnetic coupling leads to noteworthy GMR [77]. Subtle structural modifications of magnetic layers can also result in considerably altered properties and ultimately no GMR or enhanced GMR response [78]. In some cases, changes in the magnetic phase can be obtained using light, such as in [79] (Cd,Mn)Te/(Cd,Zn,Mg)Te or [80] (In,Mn)As/(Al,Ga)Sb. To obtain GMR, the electrical resistance needs to be significantly reduced by a magnetic field that induces parallel magnetization alignment in multilayers that would otherwise contain randomly oriented magnetization vectors or be spontaneously antiferromagnetic. It was observed that in layered structures of permalloy and Cu, the intermixing of the two at the interface reduces GMR [81]. On the other hand, an inverse GMR effect was detected in Fe/Cu and Cr multilayer structures [82], with thin Cr layers intercalated into Fe.

4.21 Ferromagnetic/Nonmagnetic Interfaces

Spin-dependent conductivity in GMR multilayers is significantly influenced by scattering at the interfaces between ferromagnetic and nonmagnetic layers, as predicted by ab initio calculations based on spin-dependent energy bands [83], and through a variety of experiments. This fact is not surprising given the dissipative nature of an effect based on selective passage for one spin component of the electronic current density. Less scattering and therefore less dissipation of energy occurs when the magnetization vectors of the layers are aligned, allowing the favored spin component to carry electric current with lower resistivity, shunting out the more strongly scattered component. Additionally, magnetoresistance is always reduced by spin-flip scattering of both "spin-down" and "spin-up" electrons.

For the spin-carrying electron to make it through a layer without spin-flip scattering, the thickness of the layer must be smaller than the mean free path for spin-flip. Demagnetizing fields associated with the magnetic moments of the ferromagnetic layers become more significant as the volume is increased, fact that adds to the necessity of keeping the layers thin. However, progressive reduction of size for the devices containing such ferromagnetic layers increases the magnetic field requirements for changing their magnetic state. This will eventually limit the smallest fields that can be detected with these devices, as well as the packing density of neighboring devices.

4.22 The Nonmagnetic Spacer

The oscillation of the GMR effect is usually observed in systems where the spacer layer is a nonmagnetic $3d$, $4d$, or $5d$ transition metal. Electrons are the principal charge carriers in transition metals, and according to Mott's model [84] conduction is not by d electrons, but rather due to transitions into empty d states at the Fermi level. The high density of states and low d electron velocity make it easier for the highly mobile valence sp electrons to make the transition. The effective masses of sp electrons are small compared to d electrons, resulting in a current primarily due to valence sp electrons. In Fe the d band is exchange split, making the density of states different for "spin-up" and "spin-down" electrons at the Fermi level. In contrast to ferromagnetic Fe, in nonmagnetic Cu the d bands are fully occupied and not exchange split. The valence sp bands have low density of states and high velocity electrons, resulting in a low probability of scattering and a long mean free path, making Cu a more suitable spacer layer than Fe. When the conduction electrons leave the nonmagnetic spacer layer and enter the ferromagnetic layer, they are scattered dependent on spin. Scattering in ferromagnets is spin-selective, causing the mean free path of minority spin electrons to differ from that of majority spin electrons. Thus in ferromagnetic materials, two largely independent conduction channels exist, corresponding to "spin-up" and "spin-down" electrons. Because GMR is based on spin-dependent scattering of electrons at the Fermi energy, it is an effect that can be exploited in *spintronic* devices [85].

4.23 Magnetic Tunneling

Magnetic tunneling may also be a candidate for spintronic devices, as it involves the generation, manipulation and detection of spin polarized current. Spin polarization can be expressed simply as

$$P = \frac{n_{\uparrow} - n_{\downarrow}}{n_{\uparrow} + n_{\downarrow}},$$

where n is the density of states for majority ($\uparrow$) and minority ($\downarrow$) spin-polarized electrons. Jullière [86] proposed this relationship for the magnetoresistance:

$$\mathrm{MR} = \frac{2P_1P_2}{P_1 + P_2},$$

where P_1 and P_2 are the polarizations of the two ferromagnetic layers. Clearly, the higher MR is obtained for larger spin polarizations.

In practice, techniques that measure the spin polarization of the ferromagnet's electronic states may not give results that agree, as each technique is related to the polarization in different ways, not allowing direct comparison to each other. Although the Jullière model is for tunneling MR, it is often applied to GMR structures. This is because in both cases, the ferromagnetic/spacer (barrier) layer interface electronic states influence the degree of spin polarization. It should be noted in passing that, in GMR spin valves and in MTJ electron transport occurs near the Fermi level, however it is possible for electron transport to occur at much higher energies. These are the so-called *hot electrons*, and they are utilized in spin valve transistors made of a spin valve base with two semiconductor substrates constituting the emitter and collector.

4.24 The Magnetic Tunnel Transistor

The alignment of the magnetic moments within the spin valve base determines the collector current, therefore the base is sensitive to large magnetic fields. Nevertheless, the energy of the hot electrons is limited to about 0.9 eV due to the emitter Schottky barrier height, restricting the value of the collector current to [87] ~20 nA. On the other hand, in the *magnetic tunnel transistor* the energy of the hot electrons can be adjusted by varying the bias voltage between the emitter and base across the tunnel barrier. This allows the magnetic tunnel transistor to operate over a wide energy range while obtaining high collector currents through large voltage bias [88]. The transmission of hot electrons injected into the empty states above the Fermi level is possible because of the longer attenuation lengths of majority spin hot electrons, as there is a spin dependence of the number of states available for inelastic scattering by electron–hole pair excitation [89].

Due to the nonequilibrium nature of tunneling magnetic phenomena, when the spin-polarized hot electrons are injected into the states above the Fermi level of one ferromagnetic electrode, holes are simultaneously injected into the states below the Fermi level of the counterelectrode, also by a tunneling process. The nonequilibrium holes scatter quasielastically in the first few nanometers of the material due to spin-wave emission, while at larger depth into the layer the attenuation is dominated by inelastic scattering by electron–hole pair excitations, but with opposite spin-asymmetry. In a p-type magnetic tunnel transistor, after spin-dependent transmission through the ferromagnetic base, holes are collected in the valence band of the p-type semiconductor, provided their energy and momentum allow them to overcome the Schottky barrier [90]. For *hot holes*, a spin valve effect of 130% and attenuation lengths as short as 0.6 nm have been reported [91].

4.25 Some Special Types of Ferromagnets

Recording of information can be done using electron spin, whereas traditionally, information processing and communication rely on electron charge. The question arises whether charge and spin of electrons can both be utilized at the same time to broaden device applications. Nonmagnetic materials, such as GaAs or similar III–V semiconductors, are currently employed in devices that use charge. Apparently, they can be made magnetic if high concentrations of magnetic elements are introduced [92], although this is sometimes being disputed at magnetic semiconductor conferences. Semiconductors are particularly preferred for these types of experiments, as they have the ability to be doped with impurities and thereby change their properties [93]. If these impurities are magnetic, molecular beam epitaxial growth is employed under low temperature equilibrium conditions to enhance their solubility [94]. The result is a *diluted magnetic semiconductor*, for instance $Ga_{0.947}Mn_{0.053}$ that has a high magnetic transition temperature (~110 K) which corresponds well with the calculated transition temperature based on the Ruderman–Kittel–Kasuya–Yosida (RKKY) interaction [95].

The advantage of diluted magnetic semiconductor materials is that ferromagnetism and semiconducting characteristics coexist [96]. Many effects are attributed to the exchange interaction between itinerant holes and magnetic spins [97, 98]. For (Ga,Mn)As it is assumed that holes play a critical role in the magnetic coupling between ferromagnetic semiconductors/nonmagnetic layered structures [99], whereas exchange interactions between the electrons in the semiconducting band and the localized electrons at the magnetic ions lead to interesting electronic, optical and magnetic properties [100]. Furthermore, computations show that strain has a strong influence on the valence subbands [100], therefore it may be possible to use strain engineering to control the magnetic properties resulting from the hole-mediated exchange.

In (Ga,Mn)As, the ferromagnetic double exchange between nearby pairs of Mn ions is stronger than the antiferromagnetic superexchange. Conversely, the latter remains significant in p-(Zn,Mn)Te due to the lack of electrical participation of Mn atoms in II–VI compounds [100]. Also, (Ga,Mn)As films display a considerable magnetic anisotropy because of the interaction between spin and orbital degrees of freedom of the magnetic electrons [100]. In the case of (Ga,Mn)Sb with a few percent of Mn, a pronounced anomalous Hall effect and a negative magnetoresistance below 50 K are obtained, indicating the formation of a ferromagnetic semiconductor where the Mn atoms are incorporated into the GaSb host [101]. On the other hand, p-type (Ga,Mn)N is supposed to have a Curie temperature above room temperature [102], which would add more capabilities to GaN-based structures, already in use in photonics and high power electronics. According to theoretical [103] and experimental [104, 105] investigations, the n-type (Ga,Mn)N does not display ferromagnetism above 1 K.

4.26 Colossal Magnetoresistance

Half-metallic ferromagnets are promising materials that show potential for GMR structures, as they have an absence of either majority or minority states at the Fermi energy, so that the current shows a high degree of spin polarization. For example, Fe_3O_4 with its high Curie temperature [106] (858K), [107] CrO_2, Heusler alloys, and manganites are considered to be half-metallic ferromagnets. Doped manganite perovskites such as for instance, $La_{1-x}A_xMnO_3$ have also been under study in recent years owing to the fact that they display an effect termed *colossal magnetoresistance* (CMR).

CMR is quite different from the GMR of magnetic multilayers, nevertheless the origin of the effect is still being debated. A variety of mechanisms have been proposed as an explanation, such as electron–phonon coupling [108], double exchange [109], electron–magnon interactions [110], or phase- and charge-segregation [111]. The fact remains that significant changes in the magnetoresistance of the manganite perovskites are observed at very large fields and low temperatures. Below 140 K, the parent compound $LaMnO_3$ displays an antiferromagnetic structure, while the magnetic coupling between the Mn^{3+} ions is mediated by superexchange interactions.

Particularly interesting properties are obtained when the hole doping is increased by cationic species, resulting in compounds such as $La_{0.8}A_{0.2}MnO_3$ (A = Ca, Na, Sr) which show extreme magnetoresistance values at room temperature and a high, controllable insulator–metal transition temperature well over the paramagnetic–ferromagnetic transition temperature [112]. Introducing Sr at Ca-sites sharpens the transition from metal to insulator [113], whereas hole doping with [114] Pb leads to a completely stable ferromagnetic character. On the other hand, Ho-doping with only 1 mol% raises the metal–insulator transition temperature [115]. These changes in structure and transport properties are a result of the creation of mobile Mn^{4+} species on the Mn sites [116]. If additionally, a compound such as $La_{0.7}Pb_{0.3}MnO_3$ is doped with transition metals (e.g., Fe), the latter enters the samples as Fe^{3+}, giving rise to an antiferromagnetic coupling between Fe and Mn, and fine-tuning the double exchange mechanism [117].

A significant challenge for manganite perovskites remains the fabrication of qualitatively good materials for applications, as researchers find differences between single crystal and polycrystalline specimens [118, 119]. The microstructure is to some extent responsible for these differences [120], leading to freezing of magnetic domains in partially melted samples [121], or to variations in CMR depending on the connectivity between grains [122]. Strain induced by lattice mismatch between films and substrate has an effect on the electronic phase separation [123], influencing the charged ordered state [124]. Adding metal ions improves physical properties, especially around room temperature [125]. By changing the doping level (La/Ba ratio) and adding metal ions such as Ag in $La_{1-x}Ba_xMnO_3$ thin films grown by pulsed-laser deposition, remarkably sharp metal–insulator transitions can be obtained [126]. Sintering and partial melting seem to open up new percolating conduction channels between grains, helping the ordering of Mn spins and restoring the ferromagnetic state at elevated temperatures [127].

4.27 CPP Geometry Preferred in Sensors

A variety of studies [128–131] have come to the conclusion that the CPP geometry is the most advantageous for magnetic spin valves, leading to maximum GMR performance. This is because in the CPP geometry the current flows perpendicular to the multilayers and electrons experience spin-dependent scattering at the interfaces [132]. A quantity termed *spin diffusion length* plays a crucial role in perpendicular spin transport. A short spin diffusion length results in reduced GMR [133]. The Valet–Fert model [134, 135] uses the spin diffusion length to describe CPP GMR transport in metallic ferromagnetic–normal metal–ferromagnetic multilayers. The model takes into account dissimilar scattering rates for majority and minority spins in the two channels displaying different resistivities.

When the spins flip from "up" to "down" and conversely, an electrochemical potential difference arises, similar to the charge accumulation effect at the interface between a normal metal and a superconductor. Thus, a spin accumulation builds up at the interface while a perpendicular current flows through it. Johnson was the first to calculate the spin accumulation in ferromagnetic–normal metal–ferromagnetic configurations where the thickness of the nonmagnetic metal was much longer than the spin-diffusion length [136]. On the other hand, when the spacer layer is thin and the spin diffusion length longer than the thickness of the ferromagnetic layer, the Valet–Fert model is reduced to the "two-current representation" [137]. In this simplified version, the GMR of the CPP spin valve with antiparallel coupled pinned and reference ferromagnetic layers increases with the thickness of the latter. However, too much thickness can be detrimentary to the saturation field value that needs to be sufficiently high for sensor stability.

A challenge for measuring GMR in the CPP geometry is the small resistance of the samples making them difficult to characterize. The samples must have a surface area large enough for the magnetic layers to retain the properties characteristic for large thin films. A solution is to fabricate the layers in the form of nanowires, in an alternating magnetic and nonmagnetic layer pattern such as Co/Cu [138,139]. The large aspect ratios of nanowire length vs. diameter allow for easy magnetotransport measurements. However, ill defined current directions can lead to small GMR values of for instance [140], only 6% at room temperature reaching 12% at 5 K. In the CIP geometry, due to the current flowing in the sample plane and hence through the spacer or individual magnetic layers [141], the GMR response is even more reduced. Therefore, the CPP geometry is still preferred where improved designs can lead to low resistance-area products for the spin valves, combined with high GMR values [142, 143], making it feasible to fabricate low-resistance, nanometer-sized devices for high data rate magnetic recording applications.

4.28 Spin Valves in Commercial Applications

The sensing heads in disk drives based on GMR can detect much smaller magnetic fields than earlier generation technologies. Not only can the density of bits of information on the disk be increased, but also reading the information is now faster. Other interesting applications of spin valves, such as the bipolar spin switch [144], the micromagnetometer [145], or the micro-SQUID [146], have also emerged in the previous decade, and still more are yet to come. For instance, vibrations of several tens of Å of microelectromechanical systems microbridges have been successfully measured using a $10 \times 2\,\mu m^2$ spin valve sensor displaying a 6.5% GMR [147]. The spin valve sensor is made out of a top-pinned spin valve structure Ta/NiFe/CoFe/Cu/CoFe/MnIr/Ta using ion beam deposition on a glass substrate. Prepatterning is done by photolithography and ion-beam milling.

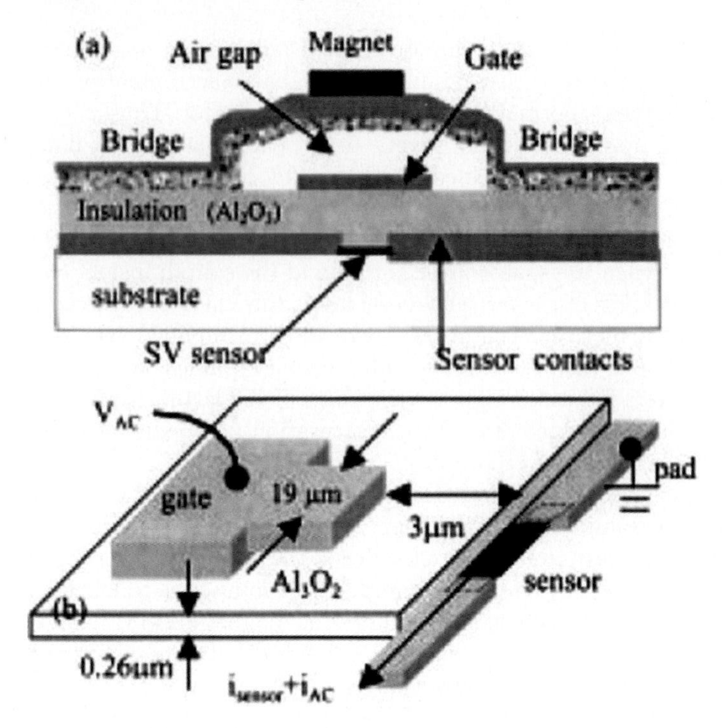

Fig. 4.4. (a) An integrated spin valve sensor and MEMS bridge with micromagnet. **(b)** The gate, spin valve sensor, and lead environment (reprinted from [147] (copyright 2002) with permission from the IEEE)

In this design, the sensor is placed 3 μm away and 2.6 μm below the central point of a Si:H/Al bridge with a 1 μm air gap (Fig. 4.4)), while it detects the fringe field (30–40 Oe) created by an rf-magnetron sputtered $Co_{78}Pt_{22}$ micromagnet deposited on top of the bridge. Magnet saturation occurs transverse to the bridge length, creating a transverse fringe field in the sensor. The movement of the bridge is controlled by an applied voltage on the gate. Further improvements in the design and experimental conditions led to reported bridge deflections up to [148] 0.23 μm. The whole device could be part of a membrane/microbridge/cantilever system, considering that microcantilevers are increasingly being used in biochip applications as detectors or actuators. The deflection of these microbridges can be determined with nanometer resolution due to the extraordinary sensitivity of spin valves to changes in magnetic fields.

From rotational speed control devices (for ABS systems) [149], to high current monitoring devices for power lines [150, 151] or positioning control devices in robotic systems [152], spin valves have been incorporated reliably wherever sensitivity to magnetic fields needs to be quantified.

References

1. M.N. Baibich, J.M. Broto, A. Fert, F. Nguyen van Dau, F. Petroff, Phys. Rev. Lett. **61**(21), 2472 (1988)
2. G. Binasch, P. Grunberg, F. Saurenbach, W. Zinn, Phys. Rev. B **39**, 4828 (1989)
3. P. Grünberg, R. Schreiber, Y. Pang, M.B. Brodsky, H. Sowers, Phys. Rev. Lett. **57**(19), 2442 (1986)
4. S.S.P. Parkin, in *Ultrathin Magnetic Structures*, ed. by B. Heinrich, J.A.C. Bland, vol. II (Springer, Berlin, 1994) p. 148
5. A. Fert, I.A. Campbell, J. Phys. F **6**, 849 (1976)
6. V.S. Gornakov, V.I. Nikitenko, W.F. Egelhoff, R.D. McMichael, A.J. Shapiro, R.D. Shull, J. Appl. Phys. **91**(10), 8272 (2002)
7. S.S.P. Parkin, D. Mauri, Phys. Rev. B **44**, 7131 (1991)
8. L. Geppert (ed.), *A Giant Leap for Disk Drives*, IEEE Spectrum **3**, 24 (1998)
9. R. Dill, R.E. Fontana, S.S.P. Parkin, C. Tsang, US Patent 5,898,548, 1999
10. S.S.P. Parkin, IBM J. Res. Develop. **42**(1), 3 (1998)
11. Z.-G. Zhu, G. Su, B. Jin, Q.-R. Zheng, Int. J. Modern Phys. B **16**(19), 2857 (2002)
12. M. Johnson, R.H. Silsbee, Phys. Rev. B **37**, 3312 (1988)
13. B.L. Johnson, R.E. Camley, Phys. Rev. B **44**(18), 9997 (1991)
14. P. Baumgart, B. Gurney, D. Wilhoit, T. Nguyen, B. Dieny, V. Speriosu, J. Appl. Phys. **69**, 4792 (1991)
15. P. Baumgart, B.A. Gurney, D.R. Wilhoit, T. Nguyen, B. Dieny, V.S. Speriosu, J. Appl. Phys. **69**(8), 4792 (1991)
16. S.S.P. Parkin, N. Morc, K.P. Roche, Phys. Rev. Lett. **64**, 2304 (1990)
17. K. Wakoh, T. Hihara, T.J. Konno, K. Sumiyama, K. Suzuki, Mat. Sci. Eng. **A217/218**, 326 (1996)
18. W.L. Brown, M.F. Jarrold, R.L. McEachern, M. Sosnowski, G. Takaoka, H. Usui, I. Yamada, Nucl. Instrum. Method. **B59/60**, 182 (1991)
19. A. Maeda, T. Tanuma, M. Kume, Mater. Sci. Eng. **A217**, 203 (1996)
20. R.D. Shull, A.J. Shapiro, V.S. Gornakov, V.I. Nikitenko, J.S. Jiang, H. Kaper, G. Leaf, S.D. Bader, IEEE Trans. Magn. **37**(4), 2576 (2001)
21. J.S. Jiang, E.E. Fullerton, C.H. Sowers, A. Inomata, S.D. Bader, IEEE Trans. Magn. **35**(5), 3229 (1999)
22. A.J. Shapiro, V.S. Gornakov, V.I. Nikitenko, R.D. McMichael, W.F. Egelhoff, Y.W. Tahk, R.D. Shull, L. Gan, J. Magn. Magn. Mat. **240**, 70 (2002)
23. E.E. Fullerton, J.S. Jiang, M. Grimsditch, C.H. Sowers, S.D. Bader, Phys. Rev. B **58**(18), 12193 (1998)
24. W.H. Meiklejohn, C.P. Bean, Phys. Rev. B **102**, 1413 (1956)
25. C. Tsang, J. Appl. Phys. **55**, 2226 (1984)
26. J. Nogués, I.K. Schuller, J. Magn. Magn. Mater. **192**, 203 (1999)
27. A.E. Berkowitz, K. Takano, J. Magn. Magn. Mater. **200**, 552 (1999)
28. W.H. Meiklejohn, J. Appl. Phys. **33**, 1328 (1962)
29. V.S. Gornakov, V.I. Nikitenko, A.J. Shapiro, R.D. Shull, J.S. Jiang, S.D. Bader, J. Magn. Magn. Mater. **246**, 80 (2002)
30. W. Zhu, L. Seve, R. Sears, B. Sinkovic, S.S.P. Parkin, Phys. Rev. Lett. **86**, 5389 (2001)
31. H. Ohldag, A. Scholl, F. Nolting, E. Arenholz, S. Maat, A.T. Young, M. Carey, J. Stöhr, Phys. Rev. Lett. **91**(1), 017203 (2003)

32. W.J. Antel, F. Perjeru, G.R. Harp, Phys. Rev. Lett. **83**, 1439 (1999)
33. L.T. Romankiw, D.A. Thompson, US Patent 4,295,173, 13 October 1981
34. R. Bozorth, *Ferromagnetism* (Van Nostrand, Princeton, NJ, 1951)
35. P.C. Andricacos, L.T. Romankiw, *Advances in Electrochemical Science and Engineering*, ed. by H. Gerischer, C. Tobias, vol. 3 (VCH, New York, 1994)
36. D.A. Thompson, Proc. AIP Conf. Magn. Magn. Mat. **24**, 528 (1975)
37. L.T. Romankiw, US Patent 3,908,194, 23 Sept. 1975
38. L.T. Romankiw, Electrochem. Soc. Proc. **PV90-8**, 39 (1990)
39. J-W. Chang, P.C. Andricacos, B. Petek, P.L. Trouilloud, L.T. Romankiw, Electrochem. Soc. Proc. **PV98-20**, 488 (1999)
40. E.I. Cooper, C. Bonhôte, J. Heidmann, Y. Hsu, P. Kern, J.W. Lam, M. Ramasubramanian, N. Robertson, L.T. Romankiw, H. Xu, IBM J. Res. Develop. **49**(1), 103 (2005)
41. N. Robertson, H.L. Hu, C. Tsang, IEEE Trans. Magn. **33**(5), 2818 (1997)
42. J.V. Powers, L.T. Romankiw, US Patent 3,652,442, 28 March 1972
43. T. Osaka, M. Takai, K. Hayashi, Y. Sogawa, K. Ohashi, Y. Yasue, M. Saito, K. Yamada, IEEE Trans. Magn. **34**, 1432 (1994)
44. M. Ramasubramanian, J. Lam, A. Hixson-Goldsmith, A. Medina, T. Dinan, N. Robertson, T. Harris, S. Yuan, Electrochem. Soc. Proc. **PV2002-27**, 298 (2003)
45. C. Bonhôte, H. Xu, E.I. Cooper, L.T. Romankiw, Electrochem. Soc. Proc. **PV2002-27**, 319 (2003)
46. P. Kern, C. Bonhôte, L.T. Romankiw, Electrochem. Soc. Proc. **PV2002-27**, 328 (2003)
47. H. Xu, J. Heidmann, Y. Hsu, Electrochem. Soc. Proc. **PV2002-27**, 307 (2003)
48. S. Parkin, X. Jiang, C. Kaiser, A. Panchula, K. Roche, M. Samant, Proc. IEEE **91**(5), 661 (2003)
49. S. Maat, A. Zeltser, J. Li, L. Nix, B.A. Gurney, Phys. Rev. B **70**, 014434 (2004)
50. K.V. O'Donovan, J.A. Borchers, S. Maat, M.J. Carey, B.A. Gurney, J. Appl. Phys. **95**(11), 7507 (2004)
51. W.C. Cain, W.H. Meiklejohn, M.H. Kryder, J. Appl. Phys. **61**, 4170 (1982)
52. A. Scholl, J. Stohr, J. Luning, J.W. Seo, J. Fompeyrine, H. Siegwart, J.P. Locquet, F. Nolting, S. Anders, E.E. Fullerton, M.R. Scheinfein, H.A. Padmore, Science **287**, 1014 (2000)
53. M.J. Carey, A.E. Berkowitz, Appl. Phys. Lett. **60**, 3060 (1991)
54. M.J. Carey, S. Maat, P. Rice, R.F.C. Farrow, R.F. Marks, A. Kellock, P. Nguyen, B.A. Gurney, Appl. Phys. Lett. **81**(6), 1044 (2002)
55. S. Maat, M.J. Carey, E.E. Fullerton, T.X. Le, P.M. Rice, B.A. Gurney, Appl. Phys. Lett. **81**(3), 520 (2002)
56. S. Maat, J. Checkelsky, M.J. Carey, J.A. Katine, J.R. Childress, J. Appl. Phys. **98**, 113907 (2005)
57. S. Maat, M.J. Carey, J.A. Katine, J.R. Childress, J. Appl. Phys. **98**, 073905 (2005)
58. S. Colins, A. Dinia, J. Appl. Phys. **91**, 5268 (2002)
59. J.S. Moodera, L.R. Kinder, T.M. Wong, R. Meservey, Phys. Rev. Lett. **74**, 3273 (1995)
60. K.I. Bolotin, F. Kuemmeth, D.C. Ralph, Phys. Rev. Lett. **97**, 127202 (2006)
61. W.F. Egelhoff Jr., L. Gan, H. Ettedgui, Y. Kadmon, C.J. Powell, P.J. Chen, A.J. Shapiro, R.D. McMichael, J.J. Mallett, T.P. Moffat, M.D. Stiles, E.B. Svedberg, J. Magn. Magn. Mater. **287**, 496 (2005)

62. Z.K. Keane, L.H. Yu, D. Natelson, Appl. Phys. Lett. **88**, 062514 (2006)
63. T. Ozue, M. Kondo, Y. Soda, S. Fukuda, S. Onodera, T. Kawana, IEEE Trans. Magn. **38**(1), 136 (2002)
64. S.S.P. Parkin, Z.G. Li, D.J. Smith, Appl. Phys. Lett. **58**, 2710 (1991)
65. J. Inoue, J. Magn. Magn. Mat. **164**, 273 (1996)
66. B. Dieny, Europhys. Lett. **17**, 261 (1992)
67. D.M. Edwards, R.B. Muniz, J. Mathon, IEEE Trans. Magn. **27**, 3548 (1991)
68. B.A. Gurney, V.S. Speriosu, J.P. Nozieres, H. Lefakis, D.R. Wilhoit, O.U. Need, Phys. Rev. Lett. **71**, 4023 (1993)
69. T.D. Boone, L. Folks, J.A. Katine, S. Maat, E. Marinero, S. Nicoletti, M. Field, G.J. Sullivan, A. Ikhlassi, B. Brar, B.A. Gurney, IEEE Trans. Magn. **42**(10), 3270 (2006)
70. J.R. Childress, M.J. Carey, M-C. Cyrille, K. Carey, N. Smith, J.A. Katine, T.D. Boone, A.A.G. Driskill-Smith, S. Maat, K. Mackay, C.H. Tsang, IEEE Trans. Magn. **42**(10), 2444 (2006)
71. A. Tanaka, Y. Shimizu, Y. Seyama, K. Nagasaka, R. Kondo, H. Oshima, S. Eguchi, H. Kanai, IEEE Trans. Magn. **38**(1), 84 (2002)
72. K. Nagasaka, Y. Seyama, L. Varga, Y. Shimizu, A. Tanaka, J. Appl. Phys. **89**, 6943 (2001)
73. O. Wunnicke, N. Papanikolaou, P.H. Dederichs, V. Drchal, J. Kudrnovsky, Phys. Rev. B **65**, 064425 (2002)
74. J.R. Childress, M.J. Carey, S.I. Kiselev, J.A. Katine, S. Maat, N. Smith, J. Appl. Phys. **99**, 08S305 (2006)
75. S.S.P. Parkin, Z.G. Li, D.J. Smith, Appl. Phys. Lett. **58**, 2710 (1991)
76. K. Liu, S.M. Zhou, C.L. Chien, V.I. Nikitenko, V.S. Gornakov, A.J. Shapiro, R.D. Shull, J. Appl. Phys. **87**(9), 5052 (2000)
77. S.S.P. Parkin, Phys. Rev. Lett. **67**, 3598 (1991)
78. F.J. Himpsel, T.A. Jung, P.F. Seidler, IBM J. Res. Develop. **42**(10), 33 (1998)
79. A. Haury, A. Wasiela, A. Arnoult, J. Cibert, S. Tatarenko, T. Dietl, Y. Merle d'Aubigné, Phys. Rev. Lett. **79**, 511 (1997)
80. S. Koshihara, A. Oiwa, M. Hirasawa, S. Katsumoto, Y. Iye, C. Urano, H. Takagi, H. Munekata, Phys. Rev. Lett. **78**, 4617 (1997)
81. V.S. Speriosu, J.P. Nozieres, B.A. Gurney, B. Dieny, T.C. Huang, H. Lefakis, Phys. Rev. B **47**, 11579 (1993)
82. J.M. George, L.G. Pereira, A. Barthélémy, F. Petroff, L. Steren, J.L. Duvail, A. Fert, R. Loloee, P. Holody, P.A. Schroeder, Phys. Rev. Lett. **72**, 408 (1994)
83. R.K. Nesbet, IBM J. Res. Develop. **42**(1), 53 (1998)
84. N.F. Mott, Proc. R. Soc. Lond. Ser. A **156**, 368 (1936)
85. J.F. Gregg, I. Petej, E. Jouguelet, C. Dennis, J. Phys. D: Appl. Phys. **35**, R121 (2002)
86. M. Jullière, Phys. Lett. **54**, 225 (1975)
87. D.J. Monsma, R. Vlutters, J.C. Lodder, Science **281**, 407 (1998)
88. S. van Dijken, X. Jiang, S.S.P. Parkin, Appl. Phys. Lett. **80**, 3364 (2002)
89. R. Jansen, J. Phys. D: Appl. Phys. 36, R289 (2003)
90. B.G. Park, T. Banerjee, J.C. Lodder, R. Jansen, Phys. Rev. Lett. **97**, 137205 (2006)
91. T. Banerjee, E. Hag, M.H. Siekman, J.C. Lodder, R. Jansen, Phys. Rev. Lett. **94**, 027204 (2005)
92. H. Ohno, Science **281**(5379), 951 (1998)

93. H. Ohno, H. Munekata, T. Penny, S. von Molnár, L.L. Chang, Phys. Rev. Lett. **68**, 2664 (1992)
94. A. Oiwa, A. Endo, S. Katsumoto, Y. Iye, H. Ohno, H. Munekata, Phys. Rev. B **59**, 5826 (1999)
95. F. Matsukura, H. Ohno, A. Shen, Y. Sugawara, Phys. Rev. B **57**(4), R2037 (1998)
96. H. Munekata, A. Zaslavsky, P. Fumagalli, R.J. Gambino, Appl. Phys. Lett. **63**, 2929 (1999)
97. T. Adhikari, S. Basu, Jpn. J. Appl. Phys. **33**(1), 4581 (1994)
98. H. Ohno, A. Shen, F. Matsukura, A. Oiwa, A. Endo, S. Katsumoto, Y. Iye, Appl. Phys. Lett. **69**, 363 (1996)
99. T. Dietl, H. Ohno, F. Matsukura, Phys. Rev. B **63**, 195205 (2001)
100. S. Koshihara, A. Oiwa, S. Katsumoto, Y. Iye, C. Urano, H. Takagi, H. Munekata, Phys. Rev. Lett. **78**, 4617 (1997)
101. F. Matsukura, E. Abe, H. Ohno, J. Appl. Phys. **87**(9), 6442 (2000)
102. M.L. Reed, N.A. El-Masry, H.H. Stadelmaier, M.K. Ritums, M.J. Reed, C.A. Parker, J.C. Roberts, S.M. Bedair, Appl. Phys. Lett. **79**(21), 3473 (2001)
103. T. Dietl, A. Haury, Y. Merle d'Aubigné, Phys. Rev. B **55**, R3347 (1997)
104. S. von Molnár, H. Munekata, H. Ohno, L.L. Chang, J. Magn. Magn. Mater. **93**, 356 (1991)
105. Y. Satoh, D. Okazawa, A. Nagashima, J. Yoshino, Physica E **10**, 196 (2001)
106. A. Yanase, K. Siratori, J. Phys. Soc. Jpn. **53**, 312 (1984)
107. K.P. Kämper, W. Schmitt, G. Güntherodt, R.J. Gambino, R. Ruf, Phys. Rev. Lett. **59**, 2788 (1987)
108. A.J. Milli, P.B. Littlewood, B.I. Shraiman, Phys. Rev. Lett. **74**, 5144 (1995)
109. K. Kubo, N. Ohata, J. Phys. Soc. Jpn. **33**, 21 (1972)
110. E.L. Nagaev, Phys. Rev. **346**, 387 (2001)
111. A. Moreo, S. Yuunoki, E. Dagotto, Science **283**, 2034 (1999)
112. X. Zhu, Y. Sun, J. Dai, W. Song, J. Phys. D: Appl. Phys. **39**, 2654 (2006)
113. A.K. Pradhan, Y. Feng, B.K. Roul, D.R. Sahu, Appl. Phys. Lett. **79**(4), 506 (2001)
114. C.W. Searle, S.T. Wang, Can. J. Phys. **48**, 2023 (1969)
115. A.K. Pradhan, B.K. Roul, Y. Feng, Y. Wu, S. Mohanty, D.R. Sahu, P. Dutta, Appl. Phys. Lett. **78**(11), 1598 (2001)
116. J. Gutierrez, F.J. Bermejo, N. Veglio, J.M. Barandiaran, P. Romano, C. Mondelli, M.A. González, A.P. Murani, J. Phys.: Condens. Matter **18**, 9951 (2006)
117. J. Gutierrez, A. Peña, J.M. Barandiaran, T. Hernandez, L. Lezama, M. Insausti, T. Rojo, Phys. Rev. B **61**, 9028 (2000)
118. H.Y. Hwang, S.-W. Cheong, N.P. Ong, B. Batlogg, Phys. Rev. Lett. **77**, 2041 (1996)
119. K.M. Krishnan, A.R. Modak, C.A. Lucas, R. Michel, H.B. Cherry, J. Appl. Phys. **79**, 5169 (1996)
120. R. Shreekala, M. Rajeswari, K. Ghosh, A. Goyal, J.Y. Gu, C. Kwon, Z. Trajanovic, T. Boettcher, R.L. Greene, R. Ramesh, T. Venkatesan, Appl. Phys. Lett. **71**, 282 (1997)
121. X.L. Wang, J. Horvat, H.K. Liu, S.X. Dou, Solid State Commun. **108**, 661 (1998)
122. A. de Andrés, M. García-Hernández, J.L. Mártinez, C. Prieto, Appl. Phys. Lett. **73**, 999 (1998)

123. M. Uehara, S. Mori, C.H. Chen, S.-W. Cheong, Nature **399**, 560 (1999)
124. Y. Ogimoto, M. Izumi, T. Manako, T. Kimura, Y. Tomioka, M. Kawasaki, Y. Tokura, Appl. Phys. Lett. **78**, 3505 (2001)
125. R. Shreekala, M. Rajeswari, S.P. Pai, S.E. Lofland, S.B. Ogale, S.M. Bhagat, M.J. Downes, R.L. Greene, R. Ramesh, T. Venkatesan, Appl. Phys. Lett. **74**, 2857 (1999)
126. A.K. Pradhan, D.R. Sahu, B.K. Roul, Y. Feng, Appl. Phys. Lett. **81**(19), 3597 (2002)
127. A.K. Pradhan, B.K. Roul, J.G. Wen, Z.F. Ren, M. Muralidhar, P. Dutta, D.R. Sahu, S. Mohanty, P.K. Patro, Appl. Phys. Lett. **76**(6), 763 (2000)
128. H.E. Camblong, S. Zhang, P. Levy, Phys. Rev. B **47**, 4735 (1993)
129. M.A.M. Gijs, G.E. Bauer, Adv. Phys. **46**, 285 (1997)
130. W.P. Pratt Jr., S.F. Lee, J.M. Slaughter, R. Loloee, P.A. Schroeder, J. Bass, Phys. Rev. Lett. **66**, 3060 (1991)
131. S.D. Steenwyck, S.Y. Hsu, R. Loloee, J. Bass, W.P. Pratt Jr., J. Magn. Magn. Mater. **170**, L1 (1997)
132. J.-Ph. Ansermet, J. Phys.: Condens. Matter. **10**, 6027 (1998)
133. A. Azizi, S.M. Thompson, K. Ounadjela, J. Gregg, P. Vennegues, A. Dinia, J. Arabski, C. Fermon, J. Magn. Magn. Mater. **148**, 313 (1995)
134. T. Valet, A. Fert, Phys. Rev. B **48**, 7099 (1993)
135. A. Fert, S. Lee, Phys. Rev. B **53**, 6554 (1996)
136. M. Johnson, Phys. Rev. Lett. **70**, 2142 (1993)
137. Q. Yang, P. Holody, S.F. Lee, L.L. Henry, R. Loloee, P.A. Schroeder, W.P. Pratt, J. Bass, Phys. Rev. Lett. **72**, 3274 (1994)
138. A. Blondel, J.Ph. Meier, B. Doudin, J.-Ph. Ansermet, Appl. Phys. Lett. **66**, 3019 (1994)
139. L. Piraux, J.M. George, J.F. Despres, C. Leroy, E. Ferain, R. Legras, K. Ounadjela, A. Fert, Appl. Phys. Lett. **65**(19), 2484 (1994)
140. A. Chalastaras, L.M. Malkinski, J.-S. Jung, S.-L. Oh, J.-K. Lee, C.A. Ventrice Jr., V. Golub, G. Taylor, IEEE Trans. Magn. **40**(4), 2257 (2004)
141. B. Dieny, V.S. Speriosu, S.S.P. Parkin, B.A. Gurney, D.R. Wilhoit, D. Mauri, Phys. Rev. B **43**(1), 1297 (1991)
142. W.P. Pratt, S.F. Lee, P. Holody, Q. Yang, R. Loloee, J. Bass, P.A. Schroeder, Phys. Rev. B **51**, 3226 (1995)
143. S.D. Steenwyck, S.Y. Hsu, R. Loloee, J. Bass, W.P. Pratt, J. Magn. Magn. Mater. **170**, L1 (1997)
144. M. Johnson, Mater. Sci. Eng. B **31**, 199 (1995)
145. V. Cros, S.F. Lee, G. Faini, A. Cornette, A. Hamzic, A. Fert, J. Magn. Magn. Mater. **165**, 512 (1997)
146. C. Chapelier, M. El Khatib, P. Perrier, A. Benoit, D. Mailly, in *Superconducting Devices and Their Applications*, ed. by H. Koch, H. Lubbig (Springer, Berlin, 1991), p. 286
147. H. Li, J. Gaspar, P.P. Freitas, V. Chu, J.P. Conde, IEEE Trans. Magn. **38**(5), 3371 (2002)
148. H. Li, M. Boucinha, P.P. Freitas, J. Gaspar, V. Chu, J.P. Conde, J. Appl. Phys. **91**(10), 7774 (2002)
149. J. Daughton, J. Brown, E. Shen, R. Beech, A. Pohm, W. Kud, IEEE Trans. Magn. **30**, 4608 (1994)

150. J.K. Spong, V.S. Speriosu, R.E. Fontana, M.M. Dovek, T.L. Hylton, IEEE Trans. Magn. **32**(2), 366 (1996)
151. N. Sugawara, M. Takiguchi, T. Yaoi, N. Negoro, K. Kagawa, A. Okabe, K. Hayashi, H. Kano, Appl. Phys. Lett. **70**(4), 523 (1997)
152. P.P. Freitas, F. Silva, N.J. Oliveira, L.V. Melo, L. Costa, N. Almeida, Sens. Actuators **81**, 2 (2000)

5

Some Basic Spintronics Concepts

Summary. New and intriguing devices for more efficient information processing have emerged in the first decade of the new millenium. An elusive property of electrons known as *spin* may just be the next entity for data encoding. This connection between electron transport and spin plays a major role in information transport through the device. Since spin is usually associated with quantum mechanics, any devices relying on it are termed *quantum devices*. In the present chapter, a new area of electronics is discussed, an area that has seen various degrees of experimental success in proving that information can indeed be encoded, transported and stored using both electron charge and spin. When dealing with spin it is expected that tiny length scales within the device come into play, and that phenomena not generally encountered in bulk materials come to light. Furthermore, for information to be reliably encoded and transported from one part of the device to another using the spin associated with electrons, certain conditions need to be met. Aside from examining some of these key conditions, this chapter also highlights what has been done to eliminate challenges encountered in spin injection, transport and detection, as well as a few developments where the majority of research seems to be concentrated nowadays.

5.1 Encoding Information: Emergence of Spintronics

Spin allows for the differentiation between electrons [1] grouping them into two types, "spin-up" and "spin-down" depending on their $\pm 1/2$ spin projection onto a given quantization axis. In ferromagnetic metals, there is an imbalance at the Fermi level in the number of "spin-up" and "spin-down" electrons [2]. Because of this imbalance, electrons traveling from one metal ferromagnet to another through a nonmagnetic spacer carry information about the magnetization of the first ferromagnet [3].

The spin of a single electron is considered a suitable entity to encode a *qubit* [4], a coherent superposition of "spin-up" and "spin-down" states. Thus, information can be encoded via electron spin, and transported from one part of the device to another using electron current [5], if certain conditions are met for the phase coherence to be maintained. The coding can be changed by remagnetizing the metal ferromagnet containing the information. Also, the transport of information only takes place provided the device is built on a short length scale for the quantum mechanical phase coherence of the electronic wave function to be preserved [6]. The quantum mechanical phase coherence is measured by the *transverse relaxation time* T_2 [7]. Spin coherence times significantly exceed charge coherence times, making spin based quantum devices more robust and fault tolerant than their charge based counterparts [8].

Control of spin-polarized electrical conduction while maintaining phase coherence is likely to have a great impact on quantum information technology [9]. It is not surprising that a promising field termed *spin electronics* or *spintronics* has emerged [10], dealing with the active control and manipulation of spin degrees of freedom in condensed matter structures. Spintronics is based on three factors [11]: *spin injection, control of spin transport*, and *spin selective detection*, however to generate the properties that are important for these three factors certain mesomagnetic length scales need to be considered, typically [12] tens to thousands of Angstroms. Nanotechnology can nowadays construct devices on these scales [13].

5.2 Spin Injection

5.2.1 Minority vs. Majority Spin Carriers

The injection of spin from a ferromagnet into an otherwise nonmagnetic material is one of the prerequisites for any spintronic device [14]. But how do we get to these spins? According to Mott's model [15] from the 1930s, "spin-up" and "spin-down" electrons exist in two different channels (or subbands) in a ferromagnetic material. Although spin-flip scattering occurs between the two channels resulting in spins losing their initial orientation, it is often neglected given the short time scales of all other processes in the system [14]. Nonetheless, the density of states in the two spin channels is different, and the electrical current is primarily due to electrons with a lower density of states at the Fermi level. These are also known as "minority spin carriers" [16], as opposed to "majority spin carriers" corresponding to those with a higher density of states. Minority spin carriers constitute a good supply of spins when the current passes from a ferromagnet to a nonmagnetic material.

5.2.2 Spin Injection Rate

The spin injection rate at a single ferromagnetic/semiconductor heterojunction is given by the ratio of the net magnetization current j_M to electric current j_e passing through the interface under an applied voltage [17]:

$$\frac{j_\mathrm{M}}{j_\mathrm{e}} = \eta_\mathrm{M} \frac{\mu_\mathrm{B}}{e}, \tag{5.1}$$

where μ_B is the Bohr magneton. The relationship is valid under the assumption that no spin-flip scattering or no spin precession takes place at the interface [17, 18]. The degree of spin polarization of the net electron flux through the interface is described by the interfacial transport parameter [14] (or spin injection efficiency) η_M. The latter is connected to the "spin-up" and "spin-down" conductances $G_\uparrow$ and $G_\downarrow$ through [14]

$$\eta_M = \frac{G_\uparrow - G_\downarrow}{G_\uparrow + G_\downarrow}. \tag{5.2}$$

For a double ferromagnetic/semiconductor/ferromagnetic heterojunction, the spin injection efficiency η'_M is defined similarly in terms of the total spin conductances [14] $G_\uparrow{}^{tot}$ and $G_\downarrow{}^{tot}$

$$\eta'_M = \frac{G_\uparrow^{tot} - G_\downarrow^{tot}}{G_\uparrow^{tot} + G_\downarrow^{tot}}. \tag{5.3}$$

The elastic multiple scattering at the interfaces leads to "spin-up" and "spin-down" transmission probabilities $T_\uparrow$ and $T_\downarrow$ that result [14] in spin injection efficiency η'_M

$$\eta'_M = \frac{T_\uparrow - T_\downarrow}{T_\uparrow + T_\downarrow - T_\uparrow T_\downarrow}. \tag{5.4}$$

A net spin current flows through the two interfaces when $T_\uparrow \neq T_\downarrow$ [14]. Spin injection can be experimentally tested using a spin valve configuration, as reviewed briefly further below.

5.2.3 Spin Polarization and Spin Transfer

In low dimensional systems, a spin polarization of electrons can exist even in zero magnetic field [19]. For metals, the spin–orbit effect can result in an energy separation of a few mV between the spin bands. Nevertheless for low electron densities, electron–electron interactions can cause a 2D homogeneous system to become unstable to spontaneous spin polarization as a result of exchange [20]. Reports of an enhanced g-factor and anomalous spin susceptibility have indicated that *spontaneous* spin polarization does exist when the disorder is low [21].

Spin-polarized current can transfer the angular momentum from carriers to a ferromagnet where it can change the direction of magnetization, even though no external field is acting. This effect is equivalent to a *spin transfer torque* [22, 23]. The spin torque is driven by a spin-polarized current [24], and is a reaction to a common process termed *spin filtering* [25], because spin-polarized electrons interact with the ferromagnet. Spin filtering means that incoming electrons with spin components perpendicular to the magnetic moment in the ferromagnet are being filtered out. Consequently, this perpendicular spin is transferred as a torque exerted on the moment due to conservation laws that have to take place. Hence, the mechanism for spin transfer implies a spin filtering process without which the transfer would not take place. Nevertheless, the ferromagnet needs to be thin, as the conditions for this effect to occur are not realized in bulk materials [26]. Furthermore, when a phenomenon termed *spin transmission resonance* [26] takes place, the magnetic moment in the ferromagnet remains unaltered, and no spin transfer occurs. Thus, there is more than one way to control the injection of spin.

5.2.4 CPP vs. CIP Geometry

To study spin injection and detection, the so-called polarize/analyze experiments [27] are used where the first ferromagnet serves as a "polarizer," and the second as an "analyzer." A voltage jump is detected when the magnetizations change from a parallel to an antiparallel orientation. Julliére's experiment [28] reported in 1975 was the first demonstration of spin filtering without superconductors, as he employed an Fe/Ge/Co structure (at 4.2 K). In principle, in a "polarize/analyze" experiment two ferromagnets are used in two geometries while separated by a nonmagnetic spacer layer (see also Chap. 4). The current-perpendicular-to-plane (CPP) geometry (Fig. 5.1, bottom) is more frequently utilized, and most mathematical treatments refer to it [29]. On the other hand, the current-in-plane (CIP) geometry (Fig. 5.1 top) was historically employed first because it is easier to realize [30].

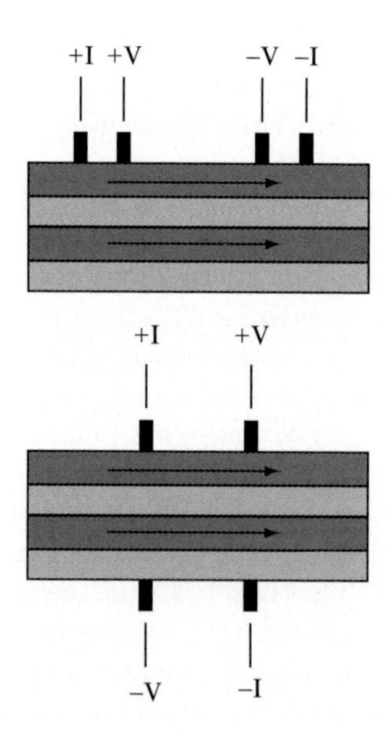

Fig. 5.1. *Top*: Current-in-plane (CIP); *bottom*: current-perpendicular-to-plane (CPP) geometry (reprinted from [38] (copyright 2002) with permission from the Institute of Physics)

5.2.5 Spin Accumulation, Spin Relaxation, and Spin Diffusion Length

In some cases, to generate spin polarization and thereby a nonequilibrium spin population [31], optical techniques have been employed involving circularly polarized photons that transfer their angular momentum to electrons [32,33]. For actual spintronics devices, electrical spin injection is preferred [34], where a magnetic electrode is connected to a specimen. A current from the electrode drives spin-polarized electrons to the sample, creating a nonequilibrium *spin accumulation* in the latter [35]. To restore equilibrium, a process of spin relaxation occurs, giving rise to an exponential spin accumulation decay from the interface. Spin relaxation ranges from picoseconds to microseconds [36,37], and takes place over a distance termed *spin diffusion length.* The spin diffusion length l_{sd} is estimated to be [38]

$$l_{\mathrm{sd}} = \sqrt{\frac{\nu_{\mathrm{F}}\tau_{\uparrow\downarrow}\lambda}{3}}, \tag{5.5}$$

where ν_{F} is the Fermi velocity, $\tau_{\uparrow\downarrow}$ is the spin-flip time, and λ is the mean free path. If any spin accumulation is to be achieved, the spin relaxation needs to be counterbalanced by a certain spin accumulation rate [39]. The latter can be given in terms of the spin density n at distance x from the interface [38]

$$n = n_0 \mathrm{e}^{-x/l_{\mathrm{sd}}}, \tag{5.6}$$

where n_0 is the spin density at the interface. After integrating to obtain the total number of spins in the accumulation, the spin accumulation rate is found to be $n_0 l_{\mathrm{sd}}/\tau_{\uparrow\downarrow}$. In the end, a total spin density of $10^{22}\,\mathrm{m}^{-3}$ is estimated, compared to a total electron density of about $10^{28}\,\mathrm{m}^{-3}$ [38]. This shows that the spin density asymmetry represents only about one part in a million of the total electron density, making it difficult to measure. Impurities shorten the mean free path and the spin-flip time, thereby reducing the spin diffusion length.

5.2.6 No Spin Accumulation in CIP Geometry

Spin accumulation cannot be achieved in a CIP geometry [38], therefore the *critical length scale* is in this case the mean free path. Nevertheless, a drop in resistance is observed even in a CIP sample when the magnetizations are aligned in the two ferromagnetic layers. This drop is attributed to different mobilities for the "spin-up" and "spin-down" electrons, and not to spin accumulation. If the nonmagnetic layer is sufficiently thin for the electrons to experience successive momentum scattering events when they reach either one of the ferromagnetic layers, a difference in "spin-up" and "spin-down" electron scattering will lead to resistance changes. Therefore, while the magnetizations in the ferromagnets are aligned one spin type is heavily scattered in both layers, leaving the other spin type relatively unscattered [29]. On the other hand, if the layers have antiparallel magnetization, neither spin type has high mobility due to heavy scattering in one ferromagnet or the other. For giant magnetoresistance [40] (GMR) to occur in CIP samples, the thickness of the spacer layer has to be less than the mean free path (20–30 Å).

5.2.7 Half-Metallic Ferromagnets

In a *half-metallic ferromagnet* [41], even in the absence of a magnetic field [42] only one spin channel has available states at the Fermi surface, making up the "majority spin carriers" [43]. The other spin band is either completely full or completely empty, while its energy level is different from that of the Fermi level [44]. This group of ferromagnets contains *Heusler alloys* [45], *ferromagnetic oxides* [46, 47], and some *diluted magnetic semiconductors* (DMS) [14].

Heusler alloys are cubic materials with four interpenetrating fcc sublattices. They are thought to include Co_2MnSi [48], NiMnSb [49], Co_2MnGe [50], and Co_2FeSi [51]. Group IV magnetic semiconductors are of particular interest due to their potential compatibility with Si-based processing technologies. However, their difficult fabrication requirements such as high doping levels and high growth temperatures may pose some challenges. When unwanted phase separations occur, the materials obtained are disordered and inhomogeneous. Usually, doping with several elements is preferred over single ones, as the dopants alter the local kinetic and strain conditions [52]. X-ray diffraction investigations revealed that strains in the film depend linearly on Co and Mn concentrations. While Co in Ge contracts the lattice, doping with Mn expands it [53]. At the same time, strain stabilizes growth in Co-rich samples due to the small lattice mismatch. There may exist an optimum doping concentration for which the lattices may no longer be mismatched. Furthermore, electric and magnetic properties of Co-rich $(Co_aMn_b)_xGe_{1-x}$ epitaxial films are controlled by the doping concentration, therefore the latter is a key factor in the fabrication of these materials. There is also a possibility that both electrons and holes are spin polarized, thereby opening new avenues for heterojunction applications [53].

Half-metallic ferromagnetic oxides consist of compounds that have itinerant electrons such as CrO_2 and Sr_2FeMoO_6, or compounds with localized electrons as is the case for Fe_3O_4. The discovery of unexpected properties in ferromagnetic oxides has sparked a general interest in these materials all over the world [53]. Of course, the type of fabrication technique is the most important factor in influencing what properties a certain compound may have.

5.2.8 Some Epitaxial Growth Techniques

Oxygen-plasma-assisted-molecular-beam-epitaxy-grown $Co_xTi_{1-x}O_{2-x}$ usually results in a high concentration of oxygen vacancies that permit thermally driven redox reactions to take place. This in turn leads to Ti being fully oxidized at the expense of Co, and the formation of Co nanoparticles. Furthermore, spectroscopic investigations reveal that O vacancies are effectively bound to substitutional Co to maintain its +2 oxidation state. Nevertheless, this material has great thermal stability as it does not change its magnetic properties when annealed in vacuum at 825 K, as opposed to pulsed-laser-deposition-grown Co : TiO_2 [54].

Recently, an advanced technique termed *combinatorial-laser-molecular-beam-epitaxy* [53] has been developed for the synthesis of a large number of samples with different compositions in a short amount of time. Compositional variations are easily implemented into high-throughput synthesis methods followed by a rapid characterization of structural, electrical, and magnetic properties. This approach allows an efficient search for advanced functional materials and identification of structure–composition–property relationships. For instance, it was applied to the investigation of new magnetic oxide semiconductors such as ZnO and TiO_2 doped with 3*d* transition metals. Generally, anatase TiO_2 is unstable and very sensitive to the choice of substrate relative to the rutile mineral phase that is easier to grow. Also, to obtain specific properties for the compound to be grown, the oxygen partial pressure and growth temperature need to be carefully chosen to better control the valence states of various transition metals.

The combinatorial-laser-molecular-beam-epitaxy approach found that none of the 3*d* transition metals doped into ZnO films resulted in ferromagnetic behavior, at least within the experimental error. On the other hand, some experimenters observed that Co-doped ZnO grown on sapphire substrates is slightly ferromagnetic up to [55] ~350 K. In this case, oxygen vacancies were noted to lead to n-type conductivity, while Co mole fractions of up to 0.35 were obtained. Co(II) was found to substitute for Zn(II) in the ZnO lattice. As far as Co-doped TiO_2 is concerned, it was observed to display the highest saturation and remanent magnetization, as well as the highest Curie temperature (~700 K) among all half-metallic ferromagnetic oxides [54]. Furthermore, a $Co_{0.06}Ti_{0.94}O_2$ film was noticed to be transparent in the visible and near-infrared regions.

5.2.9 ME Materials and Spintronics

Some magnetoelectric (ME) materials [56, 57] such as Cr_2O_3, $GaFeO_3$, $Ni_3B_7O_{13}I$, $LiMnPO_4$, and $Y_3Fe_5O_{12}$ may be useful in spintronics applications. These materials display a coexistence of spontaneous ferromagnetic (or antiferromagnetic) and ferroelectric order, where the coupling between magnetization and polarization gives rise to a ME response [58]. Antiferromagnetic order is only slightly affected by magnetic fields, however exerts a strong control of the magnetic configuration due to the exchange coupling to adjacent ferromagnetic films. This coupling can be tuned by electric fields via the ME effect. The ME influence on GMR has been demonstrated experimentally in $CrO_2/Cr_2O_3/CrO_2$ junctions, where the Cr_2O_3 film grows naturally between CrO_2 crystallites [59]. Similarly, the same research group proposes using an antiferromagnetic ME thin film as a dielectric tunnel junction between two ferromagnetic metallic layers, one soft and one hard. The ME tunnel barrier could sustain high electric fields of dielectric breakdown value while simultaneously giving rise to a significant net magnetization that would shift the magnetization curves of both ferromagnetic layers, allowing the resistance value of the device to be determined by the electric field in the ME layer [59].

5.2.10 Spontaneous Band Splitting

In DMS such as the ferromagnetic phase of (Ga,Mn)As [60] or (In,Mn)As [61], the valence band splits spontaneously into "spin-up" and "spin-down" subbands [62]. The DMS are nonmagnetic semiconductors doped with a few percent of magnetic impurities. Depending on the percentage of the dopant and the complex chemistry involved, these materials can exhibit ferromagnetism at and above room temperature. Nevertheless, what is sometimes interpreted as ferromagnetism in a DMS is actually just an effect due to a secondary magnetic impurity phase.

5.2.11 Spin Valves

For a simplistic explanation of what happens in a CPP sample, it can be assumed that the two ferromagnets are half metallic. Furthermore, it is assumed that the metals are similar, in the sense that the spin polarizations are of the same sign. In view of the density of states spin asymmetry, the first ferromagnet would supply a current with only one type of spins that accumulate in the spacer layer. If the thickness of the latter is less than the spin diffusion length, the spins will make it across the spacer into the second ferromagnet provided the magnetizations in the two ferromagnets are aligned. This is because the second ferromagnet, also a half-metal has density of states for only one spin type. Hence, spin filtering occurs. In case the magnetizations are antiparallel, the second ferromagnet will not accept the other type of spins. If the two ferromagnets in the above example are of opposite polarization, the majority spins in one ferromagnet are not a majority in the other ferromagnet. Therefore, a parallel alignment of magnetizations will lead to an increase in resistance, while the converse is true for antiparallel orientation. The overall GMR response of the system is in this case negative.

Several design and fabrication techniques [63] have been employed [64] to obtain dissimilar switching of the magnetizations in the two ferromagnetic layers [65,66], a practice common to *spin valve* [14] heads in magnetic recording. Spin valves are essentially based on how the spins scatter depending on the alignment of magnetizations, and the layer coercivities. Some aspects are also discussed in Chaps. 4 and 6.

5.2.12 Poor Injection Efficiency

When high efficiency spin injection is hindered, it is usually because of a loss of spin polarization at the interface due to spin flipping [29]. Even if the two materials are closely lattice matched, the interface is still nonideal [67]. It takes only a very small amount of ferromagnetic atoms to diffuse into the nonmagnetic side of the interface, and scatter electrons between the two spin channels [68]. This is because these diffused atoms will likely carry a local magnetic moment oriented randomly with respect to the magnetization direction, moment which interacts with that of the electrons crossing the interface [69]. Spin flipping is more pronounced if the nonmagnetic spacer has a small density of states [70]. A possible scenario for avoiding spin flipping may be the employment of a half-metallic ferromagnet due to the absence of final states for flipping [71].

Rashba [72] suggested using spin tunnel injection to circumvent the boundary condition problems related to electrochemical potential continuity at metal/semiconductor interfaces. A tunnel contact between a ferromagnet and a semiconductor is spin selective due to the spin polarization of the density of states. The barrier will lead to spin polarization of the tunnel current, and thereby support the difference in chemical potential between the "spin-up" and "spin-down" bands at the interface. However, in this case a very thin or discontinuous tunnel barrier may worsen injection efficiency.

5.2.13 Additional Layer Between Ferromagnet and Spacer

Direct injection from a "normal" ferromagnet may still be better provided that somehow spin polarization is maintained upon injection. This may be achievable if another layer of some material is introduced between the ferromagnet and the spacer. The role of this intermediate layer would be to take over the spin polarization and deliver it mostly unchanged to the spacer layer. The intermediate layer may be a magnetic or nonmagnetic tunneling barrier, or a Schottky barrier whose depletion zone is the actual intermediate layer. If the Schottky barrier is sufficiently extended, it mediates the difference in electrochemical potentials between the metal and semiconductor, improving spin injection efficiency. In case of a nonmagnetic tunneling barrier, the voltage drop across the barrier would be very large and totally control the injected current and its polarization. However, in spite of spin flipping still occurring, the injected polarization is in this case independent of the diffusion constants [73].

5.2.14 III–V Magnetic Semiconductors

High spin polarization is assumed to exist in manganite perovskites [74], and III–V ferromagnetic semiconductors like (Ga,Mn)As [75]. Mn ions are electrically neutral in II–VI compounds, but act as an acceptor in III–V semiconductors due to the large density of states in the valence band [76]. Ferromagnetic interactions between the localized spins in (Ga,Mn)As are believed to be due to holes originating from these Mn acceptors [77, 78]. In spite of the lower mobilities and shorter lifetimes as compared to electrons, holes can still provide a successful spin injection [79]. TMR values as high as 70% have been observed at 8 K for 1.6 nm AlAs barriers in GaMnAs/AlAs/GaMnAs structures [80]. Usually, III–V semiconductors are easier doped with electrically active impurities, in contrast to the II–VI compounds that can sustain greater concentrations of transition metals [81]. The magnitude of spontaneous magnetization increases with the concentration of the dopant, reaching 100% in the metallic phase [76, 82].

5.2.15 Obtaining Spin-Polarized Magnetic Semiconductors

Turning nonmagnetic semiconductors into spin-polarized ferromagnets is more difficult than it may seem at first glance, requiring among other things a certain well determined amount of magnetic elements [83]. This is because the solubility of the magnetic elements in III–V semiconductors is quite low, only $\sim 10^{18}\,\mathrm{cm}^{-3}$ [84]. Techniques such as molecular beam epitaxy or laser ablation have been used below room temperature to introduce transition metals into a semiconductor matrix without the formation of undesired phases, as for instance MnAs [85]. Some researchers found high-temperature ferromagnetism in (Ga,Mn)N [86, 87] with conclusive proof that the ferromagnetic properties are not a result of secondary magnetic phases. The GaN films were produced by metalorganic chemical vapor deposition on (0001) sapphire substrates, and doped after growth with Mn using solid state diffusion [86]. Theoretical models describing ferromagnetic behavior of Mn doped p-type semiconductors have sustained experimental findings by predicting a Curie temperature above room temperature for (Ga,Mn)N [88]. This material became of interest in photonics and high power electronics, as well as integrated systems that contain both processing of information and data storage in one unit.

To gain further insight into the properties of (Ga,Mn)As, studies related to the asymmetric scattering of spin-polarized carriers have been carried out using anomalous Hall effect data [84]. These show that all-semiconductor (Ga,Mn)As/(Al,Ga)As/(Ga,Mn)As trilayers display spin dependent scattering and GMR in a CPP or CIP configuration [89, 90] as well as a tunneling magnetoresistance (TMR) over 70% [80]. Spin injection has also been observed in spin-LEDs containing (Ga,Mn)As as an emitter [91], while a (Ga,Mn)As emitter has been employed in resonant tunneling structures [92].

5.2.16 Light vs. Electric-Field-Induced Carrier Enhancement

Ferromagnetic order in III–V diluted magnetic semiconductor heterostructures such as (In,Mn)As/GaSb was found to be induced by light [93, 94] through the enhancement of carrier-mediated ferromagnetic interactions between Mn ions. Light generates excess holes in the (In,Mn)As layer rendering it p-type. Their n-type counterparts can be paramagnetic [95]. Alternatively, it was observed that the hole concentration and ferromagnetic interactions in (In,Mn)As can be controlled by applied electric fields, a unique aspect of ferromagnetism in III–V and II–VI systems. By applying electric fields, researchers [96] managed to vary reversibly the transition temperature of hole-induced ferromagnetism in an (In,Mn)As channel part of an insulating-gate field-effect transistor. The transition temperature of (In,Mn)As is usually ∼30 K or below [97]. Nevertheless, by varying the hole concentration through applied electric fields, the ferromagnetic exchange interaction among localized Mn spins can be controlled, thereby reversibly altering ferromagnetic properties such as transition temperature. These results may be applicable to the integration of ferromagnetic semiconductors with nonmagnetic III–V semiconductor devices such as lasers.

5.2.17 Giant Planar Hall Effect

The large spin polarization of holes in (Ga,Mn)As combines its effects with the strong spin–orbit coupling in the valence band giving rise to unusual properties. Among these, researchers [98] noticed the development of a spontaneous transverse voltage in response to a longitudinal current flow in the absence of an applied field. In single domain high-quality ferromagnetic semiconductors, this effect can be ∼4 orders of magnitude stronger than in metallic ferromagnets. Furthermore, a giant planar Hall effect is observed in epitaxial (Ga,Mn)As thin films in response to an applied in plane magnetic field (Fig. 5.2). Two abrupt jumps are registered in the planar Hall resistance at distinct magnetic field values, while the longitudinal resistance also experiences small jumps. These jumps appear to be independent of sample size or geometry. Nevertheless, the smaller specimens display Barkhausen jumps (Fig. 5.2f), as the propagation of domain walls is limited by specimen geometry [99].

5.2.18 Maintaining Spin Polarization

As seen earlier, some of the problems in spintronics revolve around how to interface different materials so that spin injection is achieved, and how to maintain spin polarization at room temperature [100]. The idea of growing magnetic III–V/II–VI semiconductor material superlattices has been suggested with the intent of raising the Curie temperature [101]. Employing semiconductors is expected given that they can be processed to high purity, the equilibrium carrier densities can be varied through doping, and electronic properties are tunable by gate potentials. Furthermore, certain semiconductor features such as depletion zones, voltage blocking, diffusion currents, the tunnel effect, as well as conduction in two-dimensional electron gases [102] (2DEG) can sometimes be exploited in the design of spintronic devices.

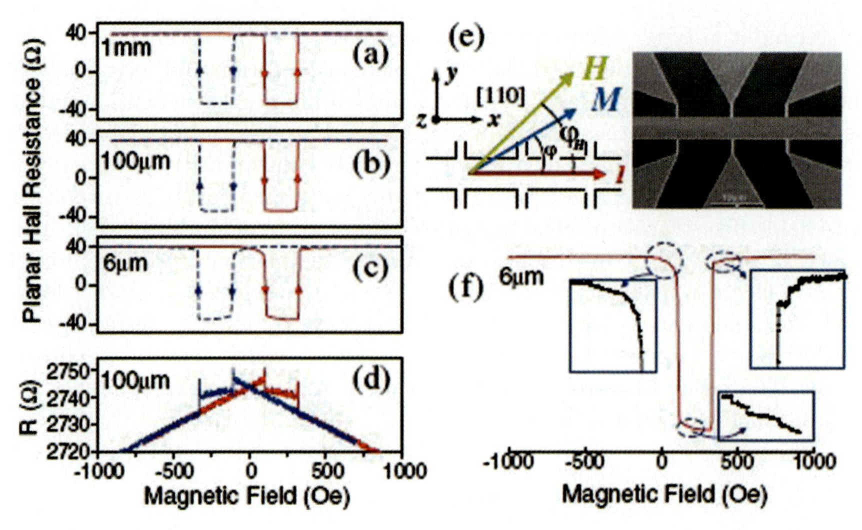

Fig. 5.2. (**a**–**c**) Planar Hall resistance in three specimens of different size obtained at 4.2 K as a function of an in-plane magnetic field at 20° fixed orientation; (**d**) field-dependent sheet resistance of the 100 μm wide Hall specimen; (**e**) relative orientations of sensing current *I*, external field *H*, and magnetization *M*. An SEM image of the 6 μm specimen is also shown; (**f**) Barkhausen jumps observable in the 6 m specimen near the resistance transitions (reprinted from [98] (copyright 2003) with permission from the American Physical Society. URL: http://link.aps.org/abstract/PRL/v90/e107201; doi:10.1103/PhysRevLett.90.107201)

In 2DEG, conduction occurs along a channel in a thin sheet of charge. 2DEGs are of interest in spintronics because they are "ballistic," as opposed to "diffusive" [103], therefore have better chances of maintaining spin polarization. However, it is unfortunate that there does not seem to be a conclusive demonstration of electrical spin injection in ohmic all-electrical ferromagnetic/semiconductor 2DEG devices [27]. The two main obstacles appear to be the local Hall effect, and the conductivity mismatch, as stated earlier [27,103]. The latter is severe for ohmic contacts between ferromagnetic metals and nonmagnetic semiconductors [104]. As already mentioned, one alternative for bypassing the problem was suggested by Rashba and consists of employing relatively high impedance magnetic tunnel junctions [72]. Nonetheless, we know that these are in a class of their own and require extensive study, as the physics of spin injection and detection with ferromagnetic/nonmagnetic tunnel junctions is rather different from that of clean ohmic ferromagnetic/nonmagnetic junctions.

Due to recent advances in high-quality magnetic semiconductors [105], and in the optimization of interfaces [106], semiconductor materials are increasingly preferred over metals. When spin-polarized carriers are injected from a magnetic semiconductor into a "normal" semiconductor, the polarization of the injected carriers depends on the magnetization direction in the magnetic semiconductor [106]. Higher percentages in injection [107] may be achieved by injection from a semiconductor, than by direct injection from metals. Nevertheless, if interface scattering is absent, the spin accumulation density may be too small in the semiconductor [108], and hence the spin-polarized current through the interface may also be too small.

5.2.19 The Future of Spin Injection

Spin injection in semiconductors differs from that in metals [109], as there is a much greater spin relaxation lifetime in semiconductors [33]. In an innovative experiment, a so-called "spin aligner" fabricated from a doped II–VI semiconductor was used to interface with a III–V compound [107]. The result was that injection efficiencies of 90% spin-polarized current were achieved, while injection took place into the nonmagnetic semiconductor device GaAs/AlGaAs used as a light-emitting diode. Spin polarization was confirmed through the circular polarization state of the emitted light [107]. In a quite different study but not of lesser significance, a ballistic *pure* spin current was injected and controlled in GaAs/AlGaAs quantum wells [110]. The current was termed pure, as it was not accompanied by a net charge current, but rather consisted only of orthogonally polarized "spin-up" and "spin-down" photons, demonstrating an alternative type of spin current.

Despite challenges posed by materials problems, the least of them being the limited functionality given by low operating temperatures, it is believed that magnetic semiconductors will play a significant role as spin injectors in the near future [62, 83]. The fact that ferromagnetism can be induced using Mn doping in zinc-blende III–V compounds or II–VI semiconductors has opened up new areas for exploration [111]. Novel physical phenomena have been discovered [112], launching the development of previously unknown spin injection devices [113].

5.3 Control of Spin Transport

Various spin decoherence mechanisms influence spin transport [114] because unlike charge, spin is a nonconserved quantity [11]. Spin-polarized electrons (or holes) maintain their polarization as long as they do not come across a magnetic impurity [115] or interact with the lattice via spin–orbit coupling. Collisions with magnons cause momentum transfer between spin channels, a process often referred to as *spin mixing* [29]. To maintain spin polarization, the interaction between spin and the transport environment needs to be controlled [116, 117].

5.3.1 The Need for Long Spin Relaxation Times

One of the challenges in spin transport is extending spin transmission over longer distances, so that spin can be more efficiently manipulated during the transfer process. Metals and inorganic semiconductors have momentum scattering lengths shorter than spin scattering lengths, leading to unwanted diffusive electron transport [118]. For this reason, organic semiconductors have recently been under increased scrutiny [119, 120] as potential replacements for metal and inorganic semiconductor spacers. It was observed that organic semiconductors display unique optical as well as spintronic properties, showing promise for a whole range of signal processing and communication applications [121]. Particularly appealing to spintronics is the weak spin–orbit interaction in organics, leading to *long spin relaxation times* [120]. Additionally, the reduced size, which can be obtained for devices incorporating organic semiconductors, allows integration into opto-spintronic chips [122].

5.3.2 Organic Semiconductor Spacers

These properties unique to organic semiconductors may be attributed to the strong intramolecular rather than intermolecular interactions. This is due to the fact that conjugated polymers are quasi-one-dimensional systems resulting in weakly screened electron–electron interactions. On the other hand, electrons and the lattice are strongly coupled; therefore, charge carriers in these materials are positive and negative polarons, rather than holes and electrons [123].

Spin valve structures with organic semiconductor spacers have been constructed in thin film [119] or nanowire form [124]. The resistance of a spin valve depends on the state of the electron spins, being controllable by an external magnetic field. Figure 5.3 depicts a hysteresis curve for Ni/Alq_3/Co trilayer

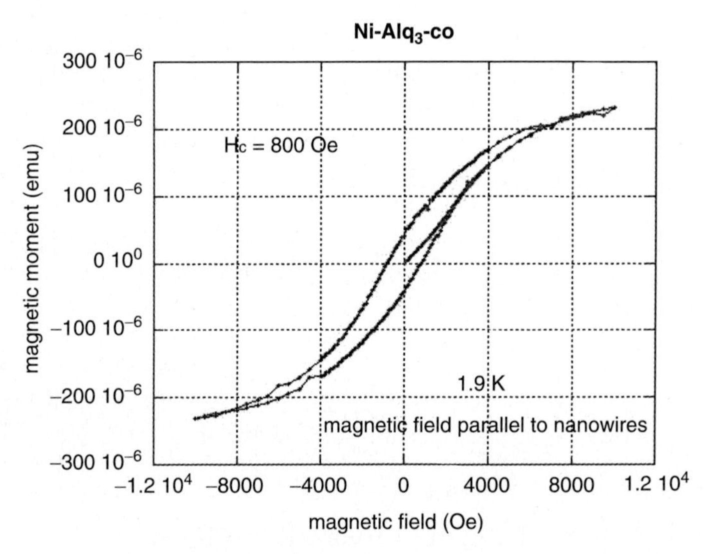

Fig. 5.3. Hysteresis curve for Ni/Alq_3/Co trilayer nanowires obtained at 1.9 K using a Physical Parameter Measurement System (Quantum Design). The trilayer nanowires were fabricated in porous alumina obtained by anodization

nanowires obtained at 1.9 K, showing that these structures are responsive to magnetic fields. When in thin film form, spin valves with organic semiconductor spacers have displayed a GMR as high as 40% at low temperatures [119]. These encouraging results may be due in part to the $La_{0.67}Sr_{0.33}MnO_3$ that is believed to be a half-metallic ferromagnet [125]. As already mentioned, half-metallic ferromagnets are appealing for spintronics, as they hold promise as superior spin injectors with near-100% spin polarization [126].

5.3.3 Spin Transport in Organic Semiconductor Spin Valves

Charge injection into organic semiconductors occurs mainly through tunneling [127], thereby avoiding the conductivity mismatch problem common to ferromagnetic/inorganic semiconductor interfaces [128]. At these interfaces, spins can switch from "spin-up" to "spin-down" and conversely, giving rise to an electrochemical potential and thereby variations in conductivity [129,130].

A recent experiment [124] with spin valves composed of nanowires with organic semiconductor spacers suggested that the primary spin relaxation mechanism in the organic material is the Elliott–Yafet mode [131] for which carrier scattering and velocity changes result in spin relaxation. These results were observed in trilayer nanowire spin valves consisting of cobalt, Alq_3 (*tris*-(8-hydroxy-quinolinolato) aluminum), and nickel, fabricated in 50 nm alumina pores. The fabrication process of the trilayers involves alumina pores obtained by anodizing an electropolished 0.1 mm thick aluminum foil in 0.3 M oxalic acid using a 40 V dc voltage [132]. Pore diameter size is flexible, ultimately dependent on anodization conditions including the type of acid employed. Also, the height of the pores is adjusted by varying the anodization time, or by subsequent etching using dilute phosphoric acid [133].

The spin valve effect is noticeable on a magnetoresistance curve when the magnetoresistance reaches a high point at certain field values. For an applied magnetic field parallel to the nanowires, the measured magnetoresistance curves revealed such points between 800 and 1,800 Oe where the coercivities of nickel and cobalt are expected. The effect was visible in the nanowires only up to $\sim$100 K.

5.3.4 Nanoscale Effects at Ferromagnet/Organic Semiconductor Interface

Because of the "proximity effect" [119], a spin polarization exists at the interface between Alq_3 and the injecting ferromagnet. The spin polarization extends over a few lattice constants. Additionally, a small Schottky barrier is formed at the organic/ferromagnet interface. Therefore, the spin-polarized carriers tunnel and diffuse through the first interface into the organic material while their spin polarization decays exponentially. They finally reach the second interface through which they also tunnel, causing a current that is detected in the second ferromagnet. The change in magnetoresistance is given by the Julliére formula [28], which takes into account spin polarization at the injecting and detecting contacts. With spin polarizations in nickel and cobalt at their Fermi energies of [134] 33% and 42%, respectively, and an Alq3 layer thickness of 33 nm, the spin diffusion length was estimated to be ~4 nm, fairly independent of the Alq3 layer thickness. This value is almost an order of magnitude smaller than the spin diffusion length reported for thin film layers [119]. This implies that the spin relaxation occurs at a higher rate in nanowires. Furthermore, carrier mobility is reduced in the Alq_3 layer because of the additional Coulomb scattering caused by the charged surface states [135]. All things considered, the Elliott–Yafet [131] spin relaxation mechanism is suggested as being dominant in the Alq3 nanowires, giving rise to long [124] spin relaxation times that range from a few milliseconds to over 1 s at 1.9 K.

5.3.5 Carbon Nanotubes

With the advent of carbon nanotubes new possibilities open up for spin transmission, especially since carbon nanotubes display extremely long electron and phase scattering lengths [136]. These unique properties are partly due to the construction of carbon nanotubes. A single walled nanotube (SWNT) consists of only one graphene sheet wrapped around to form a 1–3 nm diameter cylinder. On the other hand, multiwalled nanotubes contain several concentric cylinders reaching diameters of up to 80 nm. The chirality of the wrapping in a SWNT determines whether it behaves as a semiconductor or a metal. The lightness of the carbon atom is an advantage, as it provides for reduced spin-flip scattering.

Carbon nanotubes have been incorporated in spintronic devices with the intent of studying their spin transport properties. For instance, a MWNT was grown on a Si substrate containing a thin SiO_2 layer and ferromagnetically contacted (Fig. 5.4). In spite of the long channel nanotube length of 250 nm, the injected spin remained polarized, judging by the 9% resistance change obtained at 4.2 K [137]. Spins maintained their coherence during transport through the nanotube and influenced the value of the magnetoresistance [137].

5.3.6 GMR vs. TMR

The effective resistance of the whole multilayer or trilayer structure is dependent on magnetic changes, therefore GMR [138,139] and TMR [140,141] effects play a key role in spintronics. By varying the spacer layer composition and the measurement configuration, GMR and TMR devices can be built. An example of a GMR response is shown in Fig. 5.5, where a surprisingly large spin diffusion length was measured in a 40 nm MWNT with Co contacts [142].

In TMR, an electrical current passes by quantum mechanical tunneling through an insulating tunnel barrier between two ferromagnetic layers [143]. Two-terminal spin tunneling junctions have been built exhibiting large changes in tunnel resistance, depending on the relative orientations of the magnetizations in the ferromagnetic layers [27]. The TMR effect is usually explained based on Julliére's model [28], where the rate of tunneling is

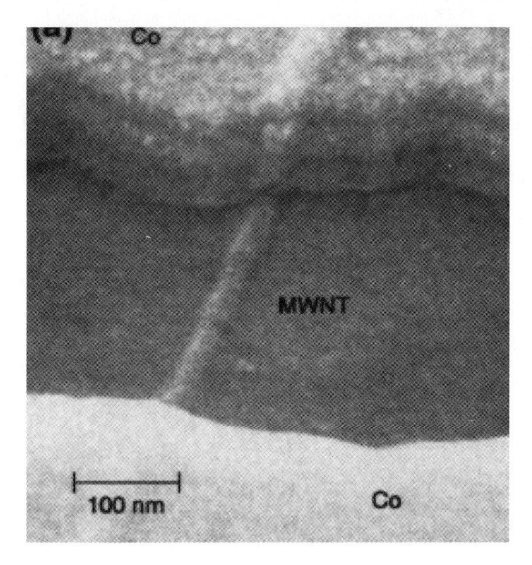

Fig. 5.4. Scanning electron microscopy image of a Co contacted MWNT. The diameter of the nanotube is 30 nm and the conducting channel length is 250 nm (reprinted from [137] (copyright 2001) with permission from the American Institute of Physics)

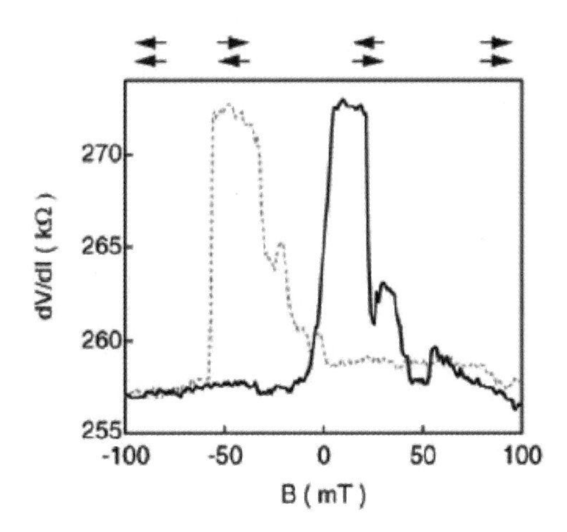

Fig. 5.5. An example of a GMR response in a 40 nm Co-contacted MWNT with a spin diffusion length of 130 nm (reprinted from [142] (copyright 2000) with permission from Elsevier)

proportional to the product of the electron densities of states at that particular energy on each side of the barrier. Electrons with majority spins tunnel if the two half-metallic ferromagnets have aligned magnetizations, rendering a low resistance for the device. During antiparallel orientation of magnetizations, the product of the initial and final densities of states is zero for either spin type, therefore no current flows [144].

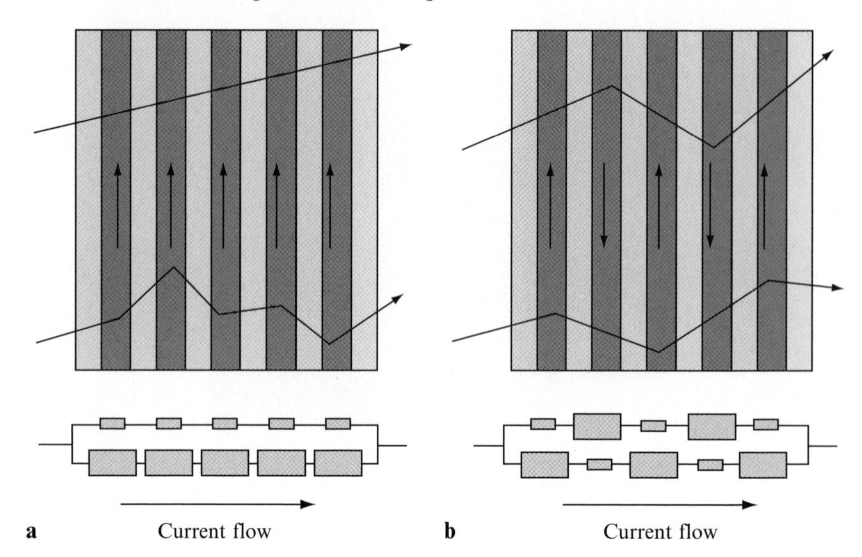

Fig. 5.6. Two resistor model provides a simple explanation for the low (**a**) and high (**b**) resistance configurations of magnetic multilayers. (**a**) The layers have parallel magnetizations allowing spins in one channel to pass through with no scattering, whereas spins in the other channel experience scattering while traversing the layers. (**b**) The magnetizations of adjacent layers are antiparallel, and spins in both channels are scattered. The resistors in each path represent the resistances the spins experience in each layer (reprinted from [38] (copyright 2002) with permission from the Institute of Physics)

5.3.7 The Parallel Resistor Model

Although crude, a parallel resistor model (Fig. 5.6) can give further insight into the mechanics of high and low resistance magnetic layer configurations [38]. As shown in Fig. 5.6, the five series resistors represent the resistances of each layer. The high and low resistances correspond to the two spin channels. With arbitrarily assigned values of 1 and 10 Ω, the parallel and antiparallel magnetization configurations have total resistances of 4.5 Ω and 13 Ω, respectively. Thus, a minimum resistance state is obtained for a parallel magnetization configuration, while antiparallel alignment results in a higher resistance. Of course, the model fails to explain what occurs at adjacent interfaces, and does not take into account spin accumulation, diffusion or relaxation.

5.3.8 Effects at Adjacent Interfaces in GMR

Some concepts have been adapted from TMR and applied to GMR because of the similarities exhibited by the two effects. To gain some insight into this issue, a very simplistic approach can be taken for GMR where spin transport is considered in a ferromagnetic multilayer stack. A spin splitting of the Fermi level is presumed in the ferromagnetic metals, causing an unequal distribution of "spin-up" and "spin-down" electron states. Additionally, if an "ohmic" nontunneling CPP configuration is assumed where Mott's model applies, then spins in one channel pass through with no scattering, while spins in the other channel are scattered at interfaces [145] when the magnetizations in all layers are parallel. Conversely, spins experience maximum scattering for the antiparallel alignment of magnetizations of neighboring magnetic layers, because spins in both channels are scattered while traversing the layers [27].

Generally, a current sourced through a ferromagnet acquires a spin polarization due to the remnant magnetization of that ferromagnet. When the current is injected into the spacer layer it induces a net polarization. The latter decays spatially with a characteristic length, the spin diffusion length mentioned earlier. Spin-polarized holes relax quickly, whereas spin-polarized electrons persist for long times [38]. If the separation between the first ferromagnet and the second is smaller than the spin diffusion length, a voltage is induced. The origin of this spin-induced voltage is the increase in "spin-down" population and decrease in "spin-up" population in the nonmagnetic spacer layer. The two spin channels are offset because of the difference experienced in their electrochemical potentials [27]. As a result, the "spin-down" electrochemical potential of the second ferromagnet must rise to match that of the "spin-downs" in the spacer layer. There is a resulting nonequilibrium magnetization in the spacer layer that depends on the partial polarization at the Fermi level, differences in Fermi velocities, and partial or inefficient spin transfer across the interface [27].

5.3.9 Scattering at Bloch Walls

Within the magnetic layers spin scattering can also occur at Bloch walls, resulting in decoherence of the electrons and wall nucleation, wich decreases the resistivity [146]. This is due to the fact that when the electron passes through the Bloch wall, the exchange field acting on the electron spin changes its orientation, an effect known in magnetic resonance spectroscopy as "adiabatic passage." The passage can be "fast" if the spin precession about the applied field is fast in comparison to the angular velocity of the rotation of the magnetic field, meaning that the spin precession follows the magnetic field in its rotation. Nevertheless, the spin precession does not quite follow the exchange field as the electron traverses the wall, so that the spins experience a potential barrier. Furthermore, if the electron passes through magnetic multilayers, or a tunnel barrier, the effect is called "sudden passage," because the electron is suddenly subjected to another magnetic field [147]. Scattering at Bloch walls was suspected of being responsible for colossal magnetoresistance (CMR) in manganate perovskites [148].

5.3.10 Importance of Materials Choice

Both GMR and TMR depend on the comprising materials [149,150]. The combination ferromagnet/spacer (or insulator) is a determining factor in obtaining a certain GMR or TMR value [151]. It is not just up to the ferromagnet to control the spin polarization, magnetoresistance or tunneling process. Therefore, proper understanding of spintronics is only possible by taking into account the electronic structure of both ferromagnetic and nonmagnetic spacer layers, and how they interface. For instance, the Fermi surfaces of Co and Cu match well for the majority spin electrons in Co, while the match is not good for the minority spins [152]. Because of this unequal alignment of subband structures in the Co/Cu system, a spin asymmetry of the tunneling current has been observed for both layered and granular structures [153]. Although large TMR can be achieved, this system is very sensitive to the interface conditions between Co and Cu requiring smooth and sharp interfaces between the magnetic and nonmagnetic layers [154]. Poor surface quality leads to "hot spots" or "pinholes." In the former case, electrons tunnel preferentially through localized regions of low barrier thickness, whereas the latter scenario implies direct contact between the ferromagnets [155].

TMR depends on the type of insulator material, as well as on the barrier height and width [156]. Additionally, TMR varies with barrier impurities, temperature, and bias voltage [157]. This is because spin-flip scattering increases with the content of magnetic impurities in the barrier, and it is assumed that it also increases with temperature, as it is an inelastic process [158]. On the other hand, some opinions [159, 160] suggest that increases in temperature bring about a reduction in the overall magnetization in the ferromagnet because of excitations of magnons. The latter effect increases electron–magnon scattering due to the fact that magnons are *spin 1* quasiparticles, leading to electron spin flipping [161], and hence a reduced TMR response.

5.3.11 Spin Control Through Electric Fields

One additional issue that should be addressed is the one of spin control through an external field. So far, most methods proposed for manipulating electron spin involve time-dependent magnetic fields. However, electric fields would be more useful from an application point of view, and this may not be so intangible. Rashba et al. suggested that while in a semiconductor quantum well electron spins couple to an electric field because of the spin–orbit interaction [162, 163]. In this case, an external in-plane electric field as small as $0.6\,\mathrm{V\,cm^{-1}}$ could be used to efficiently manipulate electron spin. Furthermore, this coupling would be by far stronger than the coupling to an ac magnetic field [163]. Future applications will put these ideas to the test.

5.4 Spin Selective Detection

Spin S can be detected by its magnetic moment $- g\mu_B S$, where g is the electron g-factor, different from the free electron value $g_0 = 2.0023$, and μ_B is the Bohr magneton [164]. With today's electron spin resonance microscopy capability, the smallest volume element that can be imaged must contain 10^7 electron spins [165]. This does not sound encouraging for single-spin detection. Fortunately, there is another technique with much improved sensitivity. Magnetic resonance force microscopy (MRFM) has proved to have the ability to detect an individual electron spin [166], because MRFM is based on detecting a magnetic force between the spins in the sample and a ferromagnetic tip. However, the force from an electron spin is only a few attonewtons, roughly 10^6 times smaller than the forces detected by atomic force microscopy (AFM) [166].

5.4.1 Detecting Single Spins

To measure unpaired electron spins, experiments involving a custom fabricated mass-loaded silicon cantilever have been performed on silicon dioxide specimens. These ultrasensitive cantilever-based force sensors are the newest development in MRFM, and are made for the purpose of facilitating single-spin detection [167]. The custom fabricated mass-loaded silicon cantilever employed in the single-spin experiment [166] had an attached 150 nm-wide SmCo magnetic tip used to sense the force from the electron spin. For the spin measurement to occur, the vitreous silica specimens had to be irradiated with Co gamma rays. This procedure resulted in silicon dangling bonds containing unpaired electron spins. Figure 5.7 shows the configuration of the single-spin MRFM experiment.

At points in the specimen where the condition for magnetic resonance is satisfied, cyclic adiabatic inversion of the spin takes place causing a slight shift of the cantilever frequency. The shift is due to the magnetic force exerted by the spin on the tip. Unfortunately, low temperatures (1.6 K) are required to minimize force noise and reduce the spin relaxation rate [166]. Nevertheless, the MRFM technique allows imaging of spins as deep as 100 nm below the surface.

5.4.2 Detecting Spin Polarization of an Ensemble of Spins

Several methods have been developed for measuring spin polarization, including spin-polarized photoemission spectroscopy [168], or Andreev reflection [169]. The latter makes use of a superconductor connected to a spin asymmetric material. A small applied bias voltage leads to electron tunneling from the normal metal to the superconductor, and the formation of Cooper pairs which have zero spin. This is because electrons with given momentum and spin couple with electrons of opposite momentum and spin, so that when the latter

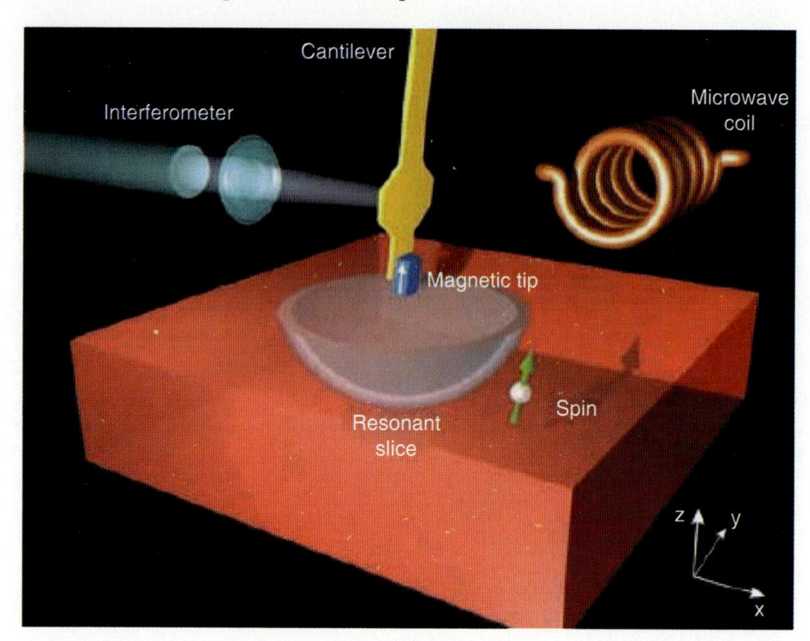

Fig. 5.7. Experimental configuration of a single-spin MRFM measurement involving an ultrasensitive silicon cantilever with a SmCo magnetic tip positioned ∼125 nm above a silica specimen containing a low density of unpaired electron spins. The "resonant slice" marks the region in the specimen where the field from the magnetic tip (plus an externally applied field) matches the condition for magnetic resonance (reprinted from [166] (copyright 2004) with permission from Nature Publishing Group)

leaves a hole is created. If the normal metal is half metallic, the necessary hole density of states is absent, therefore the Andreev reflection is suppressed. The spin polarization in the normal metal is related to the conductivity of the contact, allowing determination of the spin polarization [169].

Spin polarization can also be determined by using a reference electrode with 100% spin polarization [170]. The material under investigation constitutes the first electrode from which tunneling currents flow towards the reference electrode. Because spin polarization in the reference electrode depends not only on the electrode material itself, but also on the accompanying insulator, it is important to maintain the same magnetic metal/insulator combination in each experiment. Furthermore, good material quality and minimizing defects at interfaces can prevent spin-flip events and therefore randomization of spin orientation that could otherwise compromise the integrity of the experiment. In practical devices, not the spins themselves need to be detected, but rather changes in the spin states that lead to changes in the measured signals [171, 172].

5.4.3 The Datta and Das Spin Field Effect Transistor

In 1990, Datta and Das [173] developed the concept of a *spin field effect transistor* consisting of two ferromagnets as source (spin injector) and drain (spin detector) with parallel magnetic moments. The device was not realized experimentally, but is still useful as it illustrates some of the principles on which spintronic devices are supposed to work.

The Datta and Das transistor makes use of the Rashba effect where a stationary electric field present in the material looks partially magnetic to a relativistic traveler. This is because carrier velocities in devices are of the order of $\sim 10^6\,\mathrm{m\,s^{-1}}$, and the magnetic component of the local fields in the depletion layer, or of the crystal fields is capable of influencing the transport of the relativistic carriers. In case of the transistor, the gate voltage generates an effective magnetic field that causes the spins to precess about a precession vector due to the spin–orbit coupling. The magnetic field is tunable by the gate voltage at the top of the channel. By controlling the voltage, the precession is modified so that it leads to parallel or antiparallel spins while in transit through the channel. A large current is obtained when the spins are parallel to the magnetizations in the source and drain. The situation corresponds to the "on" state of the transistor. In the reverse situation, when the spins are antiparallel to the magnetizations in the ferromagnets, the spins are scattered and the transistor is in the "off" state.

5.4.4 The Future of Spintronics Devices

The Datta and Das [173] spin field effect transistor is an example of a device where the population and the phase of the spin of an ensemble of electrons can be controlled. Nevertheless, the prerequisites for a successful experimental implementation of spin transistors remain: (1) injection of spin-polarized current [33], (2) spin coherent propagation [174], (3) induction of controlled spin precession [175], and (4) spin selective collection [31]. In spite of these difficulties, progress is being made in all fields of spintronic devices [176,177].

The range of potential spintronic devices is starting to expand only now, and goes beyond spin field effect transistors. There still remain many challenges to be overcome before successfully fabricating commercial devices. Future developments in spintronic devices will be largely driven by advances in spintronic materials. In the end, all-semiconductor structures are expected to take over hybrid designs with metallic ferromagnets. But this will depend on the ability of producing high-quality interfaces with nonmagnetic materials.

References

1. H.C. Ohanian, Am. J. Phys. **54**(6), 500 (1986)
2. O. Pietzsch, S. Okatov, A. Kubetzka, M. Bode, S. Heinze, A. Liechtenstein, R. Wiesendanger, Phys. Rev. Lett. **96**, 237203 (2006)
3. S. Das Sarma, J. Fabian, X. Hu, I. Žuti, Superlatt. Microstruct. **27**, 289 (2000)

4. V. Privman, I.D. Wagner, G. Kventsel, Phys. Lett. A **239**, 141 (1998)
5. S. Bandyopadhyay, Phys. Rev. B **61**, 13813 (2000)
6. T. Calarco, A. Datta, P. Fedichev, E. Pazy, P. Zoller, Phys. Rev. A **68**, 012310 (2003)
7. B.E. Kane, Nature **393**, 133 (1998)
8. S. Bandyopadhyay, V.P. Roychowdhury, Superlatt. Microstruct. **22**, 411 (1997)
9. J.M. Daughton, A.V. Pohm, R.T. Fayfield, C.H. Smith, J. Phys. D **32**, R169 (1999)
10. S. Das Sarma, Am. Sci. **89**, 516 (2001)
11. S.A. Chambers, Y.K. Yoo, Mater. Res. Soc. Bull. **28**, 706 (2003)
12. A. Steane, Rep. Prog. Phys. **61**, 117 (1998)
13. C.-G. Stefanita, F. Yun, M. Namkung, H. Morkoc, S. Bandyopadhyay, in *Electrochemical Self-Assembly as a Route to Nanodevice Processing*, ed. by H.S. Nalwa. Handbook of Electrochemical Nanotechnology (American Scientific, Los Angeles 2006)
14. F. Mireles, G. Kirczenow, Europhys. Lett. **59**(1), 107 (2002)
15. N.F. Mott, Proc. R. Soc. Lond. Ser. A **156**, 368 (1936)
16. A. Barthelemy, A. Fert, Phys. Rev. B **43**, 13124 (1991)
17. M. Johnson, R.H. Silsbee, Phys. Rev. B **37**, 5326 (1988)
18. M. Johnson, Phys. Rev. B **58**, 9635 (1998)
19. A. Ghosh, C.J.B. Ford, M. Pepper, H.E. Beere, D.A. Ritchie, Phys. Rev. Lett. **92**, 116601 (2004)
20. D. Varsano, S. Moroni, G. Senatore, Europhys. Lett. **53**, 348 (2001)
21. E. Tutuc, S. Melinte, M. Shayegan, Phys. Rev. Lett. **88**, 036805 (2002)
22. X. Waintal, E.B. Myers, P.W. Brouwer, D.C. Ralph, Phys. Rev. B **62**, 12317 (2000)
23. M.D. Stiles, A. Zangwill, Phys. Rev. B **66**, 014407 (2002)
24. J.C. Slonczewski, J. Magn. Magn. Mater. **159**, L1 (1996)
25. L. Berger, Phys. Rev. B **54**, 9353 (1996)
26. W. Kim, F. Marsiglio, Phys. Rev. B **69**, 212406 (2004)
27. D.D. Awschalom, D. Loss, N. Samarth (eds.), *Semiconductor Spintronics and Quantum Computation* (Springer, Berlin, 2002)
28. M. Julliére, Phys. Lett. **54**, 225 (1975)
29. J.-Ph. Ansermet, J. Phys.: Condens. Matter **10**, 6027 (1998)
30. B. Dieny, V.S. Speriosu, S.S.P. Parkin, B.A. Gurney, D.R. Wilhoit, D. Mauri, Phys. Rev. B **43**, 1297 (1991)
31. D. Hagele, M. Oestreich, W.W. Ruhle, N. Nestle, K. Ebert, Appl. Phys. Lett. **73**, 1580 (1998)
32. D.D. Awschalom, J.M. Kikkawa, Phys. Today **52**(6), 33 (1999)
33. J.M. Kikkawa, D.D. Awschalom, Nature **397**, 139 (1999)
34. M. Johnson, R.H. Silsbee, Phys. Rev. Lett. **55**, 1790 (1985)
35. D. Chiba, M. Yamanouchi, F. Matsukura, H. Ohno, Science **301**, 943 (2003)
36. F.X. Bronold, I. Martin, A. Saxena, D.L. Smith, Phys. Rev. B **66**, 233206 (2002)
37. R.S. Britton, T. Grevatt, A. Malinowski, R.T. Harley, P. Perozzo, A.R. Cameron, A. Miller, Appl. Phys. Lett. **73**, 2140 (1998)
38. J.F. Gregg, I. Petej, E. Jouguelet, C. Dennis, J. Phys. D: Appl. Phys. **35**, R121 (2002)
39. R. de Sousa, S. Das Sarma, Phys. Rev. B **68**, 155330 (2003)

40. M.N. Baibich, J.M. Broto, A. Fert, F. Nguyen Van Dau, F. Petroff, Phys. Rev. Lett. **61**(21), 2472 (1988)
41. R.A. de Groot, F.M. Mueller, P.G. van Engen, K.H.J. Buschow, Phys. Rev. Lett. **50**, 2024 (1983)
42. W.E. Pickett, J.S. Moodera, Phys. Today **54**(5), 39 (2001)
43. R.A. de Groot, F.M. Mueller, P.G. van Engen, K.H.J. Buschow, J. Appl. Phys. **55**, 2151 (1984)
44. Y.K. Yoo, F.W. Duewer, H. Yang, Y. Dong, X.-D. Xiang, Nature **406**, 704 (2000)
45. C. Palmström, Mater. Res. Soc. Bull. **28**, 725 (2003)
46. T. Dietl, H. Ohno, F. Matsukura, J. Cibert, D. Ferrand, Science **287**, 1019 (2000)
47. K. Sato, H. Katayama-Yoshida, Jpn. J. Appl. Phys. **39**(2), L555 (2000)
48. I. Galanakis, P.H. Dederichs, N. Papanikolau, Phys. Rev. B **66**, 174429 (2002)
49. R. de Groot, F. Mueller, P. van Engen, K. Buschow, Phys. Rev. Lett. **50**, 2024 (1983)
50. S. Ishida, S. Fujii, S. Kashiwagi, S. Asano, J. Phys.: Condens. Matter **2**, 8583 (1990)
51. M. Hashimoto, J. Herfort, H-P. Schönherr, K.H. Ploog, J. Phys.: Condens. Matter **18**, 6101 (2006)
52. F. Tsui, L. He, D. Lorang, A. Fuller, Y.S. Chu, A. Tkachuk, S. Vogt, Appl. Surf. Sci. **252**(7), 2512 (2006)
53. Y. Matsumoto, H. Koinuma, T. Hasegawa, I. Takeuchi, F. Tsui, Y.K. Yoo, Mater. Res. Bull. **28**, 734 (2003)
54. S.A. Chambers, R.F.C. Farrow, Mat. Res. Bull. **28**, 729 (2003)
55. K. Ueda, H. Tabata, T. Kawai, Appl. Phys. Lett. **79**, 988 (2001)
56. I.E. Dzyaloshinskii, Sov. Phys. – JETP **10**, 628 (1960)
57. T. Katsufuji, S. Mori, M. Masaki, Y. Moritomo, N. Yamamoto, H. Takagi, Phys. Rev. B **64**, 104419 (2001)
58. T. Kimura, S. Kawamoto, I. Yamada, M. Azuma, M. Takano, Y. Tokura, Phys. Rev. B **67**, 180401 (2003)
59. Ch. Binek, B. Doudin, J. Phys.: Condens. Matter **17**, L39 (2005)
60. S.J. Potashnik, K.C. Ku, S.H. Chun, J.J. Berry, N. Samarth, P. Schiffer, Appl. Phys. Lett. **79**, 1495 (2001)
61. T. Omiya, F. Matsukura, T. Dietl, Y. Ohno, T. Sakon, M. Motokawa, H. Ohno, Physica E **7**, 976 (2000)
62. T. Dietl, H. Ohno, Mater. Res. Soc. Bull. **28**, 714 (2003)
63. A.J. Shapiro, V.S. Gornakov, V.I. Nikitenko, R.D. McMichael, W.F. Egelhoff, Y.W. Tahk, R.D. Shull, L. Gan, J. Magn. Magn. Mat. **240**, 70 (2002)
64. E.E. Fullerton, J.S. Jiang, M. Grimsditch, C.H. Sowers, S.D. Bader, Phys. Rev. B **58**(18), 12193 (1998)
65. R.D. Shull, A.J. Shapiro, V.S. Gornakov, V.I. Nikitenko, J.S. Jiang, H. Kaper, G. Leaf, S. D. Bader, IEEE Trans. Mag. **37**(4), 2576 (2001)
66. V.S. Gornakov, V.I. Nikitenko, W.F. Egelhoff, R.D. McMichael, A.J. Shapiro, R.D. Shull, J. Appl. Phys. **91**(10), 8272 (2002)
67. B.T. Jonker, Proc. IEEE **91**, 727 (2003)
68. P.M. Levy, Adv. Res. Appl. **47**, 367 (1994)
69. P.B. Allen, Solid State Commun. **102**, 127 (1997)
70. M.A.M. Gijs, E.W. Bauer, Adv. Phys. **46**, 285 (1997)

71. P.C. van Son, H. van Kempen, P. Wyder, Phys. Rev. Lett. **58**, 2271 (1987)
72. E.I. Rashba, Phys. Rev. B **62**, R16267 (2000)
73. B.T. Jonker, S.C. Erwin, A. Petrou, A.G. Petukhov, Mater. Res. Soc. Bull. **28**, 740 (2003)
74. J.-H. Park, E. Vescovo, H.-J. Kim, C. Kwon, R. Ramesh, T. Venkatesan, Phys. Rev. Lett. **81**, 1953 (1998)
75. J.G. Braden, J.S. Parker, P. Xiong, S.H. Chun, N. Samarth, Phys. Rev. Lett. **91**, 056602 (2003)
76. T. Dietl, Physica E **10**, 120 (2001)
77. H. Akai, Phys. Rev. Lett. **81**, 3002 (1998)
78. T. Dietl, H. Ohno, F. Matsukura, Phys. Rev. **63**, 195205 (2001)
79. E. Johnston-Halperin, D. Lofgreen, R.K. Kawakami, D.K. Young, L. Coldren, A.C. Gossard, D.D. Awschalom, Phys. Rev. B **65**, 041306(R) (2002)
80. M. Tanaka, Y. Higo, Phys. Rev. Lett. **87**, 026602 (2001)
81. T. Dietl, H. Ohno, Physica E **9**, 185 (2001)
82. D. Ferrand, J. Cibert, A. Wasiela, C. Bourgognon, S. Tatarenko, G. Fishman, T. Andrearczyk, J. Jaroszynski, S. Kolesnik, T. Dietl, B. Barbara, D. Dufeu, Phys. Rev. B **63**, 085201 (2001)
83. H. Ohno, Science **281**, 951 (1998)
84. H. Ohno, H. Munekata, T. Penny, S. von Molnár, L.L. Chang, Phys. Rev. Lett. **68**, 2664 (1992)
85. H. Ohno, A. Shen, F. Matsukura, A. Oiwa, A. Endo, S. Katsumoto, Y. Iye, Appl. Phys. Lett. **69**, 363 (1996)
86. M.L. Reed, N.A. El-Masry, H.H. Stadelmaier, M.K. Ritums, M.J. Reed, C.A. Parker, J.C. Roberts, S.M. Bedair, Appl. Phys. Lett. **79**, 3473 (2001)
87. H. Hori, S. Sonoda, T. Sasaki, Y. Yamamoto, S. Shimizu, K. Suga, K. Kindo, Physica B, Condens. Matter 324 (1-4), 142 (2002)
88. T. Dietl, F. Matsukura, H. Ohno, Phys. Rev. B **66**, 033203 (2002)
89. F. Matsukura, D. Chiba, Y. Ohno, T. Dietl, H. Ohno, Physica E **16**, 104 (2003)
90. D. Chiba, N. Akiba, F. Matsukura, Y. Ohno, H. Ohno, Appl. Phys. Lett. **77**, 1873 (2000)
91. M. Kohda, Y. Ohno, K. Takamura, F. Matsukura, H. Ohno, Jpn. J. Appl. Phys. **40**, L1274 (2001)
92. H. Ohno, N. Akiba, F. Matsukura, A. Shen, K. Ohtani, Y. Ohno, Appl. Phys. Lett. **73**, 363 (1998)
93. H. Munekata, T. Abe, S. Koshihara, A. Oiwa, M. Hirasawa, S. Katsumoto, Y. Iye, C. Urano, H. Takagi, J. Appl. Phys. **81**(8), 4862 (1997)
94. S. Koshihara, A. Oiwa, M. Hirasawa, S. Katsumoto, Y. Iye, C. Urano, H. Takagi, Phys. Rev. Lett. **78**(24), 4617 (1997)
95. S. von Molnár, H. Munekata, H. Ohno, L.L. Chang, J. Magn. Magn. Mater. **93**, 356 (1991)
96. H. Ohno, D. Chiba, F. Matsukura, T. Omyia, E. Abe, T. Dietl, Y. Ohno, K. Ohtani, Nature **408**, 944 (2000)
97. H. Munekata, A. Zaslavsky, P. Fumagalli, R.J. Gambino, Appl. Phys. Lett. **63**, 2929 (1993)
98. H.X. Tang, R.K. Kawakami, D.D. Awschalom, M.L. Roukes, Phys. Rev. Lett. **90**, 107201 (2003)
99. T. Ono, H. Miyajima, K. Shigeto, K. Mibu, N. Hosoito, T. Shinjo, Science **284** (5413), 468 (1999)

100. A. Shen, H. Ohno, F. Matsukura, Y. Sugawara, Y. Ohno, N. Akiba, T. Kuroiwa, Jpn. J. Appl. Phys. **36**, L73 (1997)
101. F. Matsukura, H. Ohno, A. Shen, Y. Sugawara, Phys. Rev. B **57**, R2037 (1998)
102. N. Biyikli, J. Xie, Y.-T. Moon, F. Yun, C.-G. Stefanita, S. Bandyopadhyay, H. Morkoç, I. Vurgaftman, J.R. Meyer, Appl. Phys. Lett. **88**, 142106 (2006)
103. J.M. Kikkawa, I.P. Smorchkova, N. Samarth, D.D. Awschalom, Science **277** (5330), 1284 (1997)
104. G. Schmidt, D. Ferrand, L.W. Molenkamp, A.T. Filip, B.J. van Wees, Phys. Rev. B **62**(8), R4790 (2000)
105. Y. Ohno, D.K. Young, B. Beschoten, F. Matsukura, H. Ohno, D.D. Awschalom, Nature **402**, 790 (1999)
106. B.T. Jonker, Y.D. Park, B.R. Bennett, H.D. Cheong, G. Kioseoglou, A. Petrou, Phys. Rev. B **62**(12), 8180 (2000)
107. R. Fiederling, M. Keim, G. Reuscher, W. Ossau, G. Schmidt, A. Waag, L.W. Molenkamp, Nature **402**, 787 (1999)
108. R.P. Borges, C. Dennis, J.F. Gregg, E. Jouguelet, K. Ounadjela, I. Petej, S.M. Thompson, M.J. Thornton, J. Phys. D **35**, 186 (2002)
109. S.A. Wolf, D.D. Awschalom, R.A. Buhrman, J.M. Daughton, S. von Molnár, M.L. Roukes, A.Y. Chtchelkanova, D.M. Treger, Science **294**, 1488 (2001)
110. M.J. Stevens, A.L. Smirl, R.D.R. Bhat, A. Najmaie, J.E. Sipe, H.M. van Driel, Phys. Rev. Lett. **90**(13), 136603 (2003)
111. D.K. Young, E. Johnston-Halperin, D.D. Awschalom, Y. Ohno, H. Ohno, Appl. Phys. Lett. **80**, 1598 (2002)
112. H. Boukari, P. Kossacki, M. Bertolini, D. Ferrand, J. Cibert, S. Tatarenko, A. Wasiela, J.A. Gaj, T. Dietl, Phys. Rev. Lett. **88**, 207204 (2002)
113. M. Kohda, Y. Ohno, K. Takamura, F. Matsukura, H. Ohno, Jpn. J. Appl. Phys. **40**(2), L1274 (2001)
114. D.D. Awschalom, Physica E **10**, 1 (2001)
115. J. Korringa, Can. J. Phys. **34**, 1290 (1956)
116. G. Bastard, R. Ferreira, Surf. Sci. **267**, 335 (1992)
117. G. Burkhard, H.A. Engel, D. Loss, Fortschr. Phys. **48**, 965 (2000)
118. G. Schmidt, D. Ferrand, L.W. Molenkamp, A.T. Filip, B.J. van Wees, Phys. Rev. B **62**, R4790 (2000)
119. Z.H. Xiong, D. Wu, Z.V. Vardeny, J. Shi, Nature **427**, 821 (2004)
120. V. Dediu, M. Murgia, F.C. Matacotta, C. Taliani, S. Barbanera, Solid State Commun. **122**, 181 (2002)
121. S. Forrest, P. Burrows, M. Thompson, IEEE Spectr. **37**, 29 (2000)
122. C.D. Dimitrakopoulos, P.R.L. Malenfant, Adv. Mater. **14**, 99 (2002)
123. M. Wohlgenannt, Z.V. Vardeny, J. Shi, T.L. Francis, X.M. Jiang, O. Mermer, G. Veeraraghavan, D. Wu, Z.H. Ziong, IEE Proc. Circuits Devices Syst. **152**(4), 385 (2005)
124. S. Pramanik, C.-G. Stefanita, S. Patibandla, S. Bandyopadhyay, K. Garre, N. Harth, M. Cahay, Nat. Nanotechnol. **2**, 216 (2007)
125. H.Y. Hwang, S.W. Cheong, N.P. Ong, B. Batlogg, Phys. Rev. Lett. **77**, 2041 (1996)
126. A.P. Ramirez, J. Phys.: Condens. Matter **9**, 8171 (1997)
127. H. Bassler, Polym. Adv. Technol. **9**, 402 (1998)
128. G. Schmidt, D. Ferrand, L.W. Molenkamp, A.T. Filip, B.J. van Wees, Phys. Rev. B: Condens. Matter **62**, R4790 (2000)

129. P.C. van Son, H. van Kempen, P. Wider, Phys. Rev. Lett. **58**, 2271 (1987)
130. M. Johnson, R.J. Silsbee, Phys. Rev. B **35**, 4959 (1987)
131. R.J. Elliott, Phys. Rev. **96**, 266 (1954)
132. C.-G. Stefanita, F. Yun, H. Morkoc, S. Bandyopadhyay, *Self-assembled Quantum Structures Using Porous Alumina on Silicon Substrates*, in Recent Res. Develop. Phys. vol. 5, p. 703, ISBN 81-7895-126-6, Transworld Research Network 37/661(2), Trivandrum-695 023, Kerala, India, 2004
133. C.-G. Stefanita, S. Pramanik, A. Banerjee, M. Sievert, A.A. Baski, S. Bandyopadhyay, J. Crystal Growth **268**(3–4), 342 (2004)
134. E.Y. Tsymbal, O.N. Mryasov, P.R. LeClair, J. Phys. Condens. Matter **15**, R109 (2003)
135. V. Pokalyakin, S. Tereshin, A. Varfolomeev, D. Zaretsky, A. Baranov, A. Banerjee, Y. Wang, S. Ramanathan, S. Bandyopadhyay, J. Appl. Phys. **97**, 124306 (2005)
136. R. Saito, G. Dresselhaus, M.S. Dresselhaus, *Physical Properties of Carbon Nanotubes* (Imperial College, Singapore, 1998)
137. B.W. Alphenaar, K. Tsukagoshi, M. Wagner, J. Appl. Phys. **89**(11), 6864 (2001)
138. M.A.M. Gijs, G.E.W. Bauer, Adv. Phys. **46**, 285 (1997)
139. E. Hirota, H. Sakakima, K. Inomata, *Giant Magneto-resistance Devices* (Springer, Berlin, 2002)
140. T. Miyazaki, N. Tezuka, J. Magn. Magn. Mater. **139**, L231 (1995)
141. J. Mathon, A. Umerski, Phys. Rev. B **63**, 220403 (2001)
142. B.W. Alphenaar, K. Tsukagoshi, H. Ago, Physica E **6**, 848 (2000)
143. P.M. Tedrov, R. Merservey, Phys. Rev. Lett. **26**, 192 (1971)
144. S. Maekawa, U. Gafvert, IEEE Trans. Magn. **18**, 707 (1982)
145. G. Binasch, P. Grünberg, F. Saurenbach, W. Zinn, Phys. Rev. B **39**, 4828 (1989)
146. G. Tatara, H. Fukuyama, Phys. Rev. Lett. **78**, 3773 (1997)
147. J.F. Gregg, W. Allen, S.M. Thompson, M.L. Watson, G.A. Gehring, J. Appl. Phys. **79**, 5593 (1996)
148. S.F. Zhang, Z. Yang, J. Appl. Phys. **79**, 7398 (1996)
149. S. Maekawa, S. Takahashi, H. Imamura, Mater. Sci. Eng. B **84**, 44 (2001)
150. N. Garcia, M. Munoz, Y.-W. Zhao, Phys. Rev. Lett. **82**, 2923 (1999)
151. J.M. De Teresa, A. Barthelemy, A. Fert, J-P. Contour, F. Montaigne, P. Seneor, Science **286**, 507 (1999)
152. J.E. Ortega, F.J. Himpsel, G.J. Mankey, R.F. Willis, Phys. Rev. B **47**, 1540 (1993)
153. J. Mathon, M. Villeret, R.B. Muniz, J. d'Albuquerque e Castro, D.M. Edwards, Phys. Rev. Lett. **74**, 3696 (1995)
154. P. Segovia, E.G. Michel, J.E. Ortega, Phys. Rev. Lett. **77**, 3455 (1996)
155. T. Dimopulos, V. Da-Costa, C. Tiusan, K. Ounadjela, H.A.M. van den Berg, J. Appl. Phys. **89**, 7371 (2001)
156. J.S. Moodera, L.R. Kinder, T.M. Wong, R. Meservey, Phys. Rev. Lett. **74**, 3273 (1995)
157. S.T. Chui, Phys. Rev. B **55**, 5600 (1997)
158. J. Inoue, S. Maekawa, J. Magn. Magn. Mater. **198**, 167 (1999)
159. J.S. Moodera, J. Nowak, R.J.M. van de Veerdonk, Phys. Rev. Lett. **80**, 2941 (1998)

160. A.H. MacDonald, T. Jungwirth, M. Kasner, Phys. Rev. Lett. **81**, 705 (1998)
161. S. Zhang, P.M. Levy, A.C. Marley, S.S.P. Parkin, Phys. Rev. Lett. **79**, 3744 (1997)
162. A. Bychkov, E.I. Rashba, J. Phys. C **17**, 6039 (1984)
163. E.I. Rashba, Al. L. Efros, Appl. Phys. Lett. **83**(25), 5295 (2003)
164. F. Beuneu, P. Monod, Phys. Rev. B **18**, 2422 (1978)
165. A. Blank, C.R. Dunnam, P.P. Borbat, J.H. Freed, J. Magn. Reson. **165**, 116 (2003)
166. D. Rugar, R. Budakian, H.J. Mamin, B.W. Chui, Nature **430**, 329 (2004)
167. T.D. Stowe, K. Yasumura, T.W. Kenny, D. Botkin, K. Wago, D. Rugar, Appl. Phys. Lett. **71**, 288 (1997)
168. J-H. Park, E. Vescovo, H-J. Kim, C. Kwon, R. Ramesh, T. Venkatesan, Nature **392**, 794 (1998)
169. R.J. Soulen Jr., J.M. Byers, Science 282 (5386), 85 (1998)
170. J.M. de Teresa, A. Barthelemy, A. Fert, J.P. Contour, F. Montaigne, P. Seneor, Science **286**, 507 (1999)
171. R.D.R. Bhat, J.E. Sipe, Phys. Rev. Lett. **85**, 5432 (2000)
172. B.T. Jonker, US Patent No. 5,874,749, 23 February 1999
173. S. Datta, B. Das, Appl. Phys. Lett. **56**, 665 (1990)
174. J. Nitta, F. Meijer, Y. Narita, H. Takayanagi, Physica E **6**, 318 (2000)
175. A.F. Morpurgo, J.P. Heida, T.M. Klapwijk, B.J. van Wees, G. Borghs, Phys. Rev. Lett. **80**, 1050 (1998)
176. R.I. Dzhioev, K.V. Kavokin, V.L. Korenev, M.V. Lazarev, B.Ya. Meltser, M.N. Stepanova, B.P. Zakharchenya, D. Gammon, D.S. Katzer, Phys. Rev. B **66**, 245204 (2002)
177. T. Hayashi, M. Tanaka, A. Asamitsu, J. Appl. Phys. **87**, 4673 (2000)

6

Trends in Magnetic Recording Media

Summary. Magnetic tapes have been popular for nonvolatile information storage for a long time, however the past decade has seen a significant growth of new types of memory. This is because of increasing demands for higher data storage densities that have pressured manufacturers to produce storage systems with higher capacity. To meet these demands, the granular films of the Co-based alloy were reduced in thickness and grain size, however grain size reduction below $\sim$10 nm posed problems. This is due to thermal fluctuations becoming uncontrollable, therefore leading to loss of information. The present chapter discusses the latest developments aimed at overcoming these challenges. As such, the capabilities and limitations encountered by magnetic tapes as recording media are reviewed. Additionally, new tendencies in magnetic recording media are uncovered, by examining closer discrete bit media. The latter have proven so far to have several distinct advantages over continuous media. This is in part due to their discrete bit-cells being able of maintaining a thermally stable single domain remanent state. For this reason, it appears that bit patterning the magnetic medium may be the key for next generation magnetic data storage solutions if certain challenges are overcome. As an alternative route, self-assembly may constitute the means of manufacturing future magnetic media, therefore a brief look is taken at this option.

6.1 The Popularity of Magnetic Tapes

Magnetic tapes are economical because of their reduced per-bit cost and high reliability. It is therefore not surprising that they have constituted for decades an adequate medium for archival data storage. Traditionally, magnetic recording media are made of magnetically hard layers such as *hcp* Co-based alloys containing Pt, Cr, Ta, and B sputter deposited onto an NiP-coated aluminum or glass substrate [1]. Of course, the magnetic easy axis of Co coincides with its *hcp* *c*-axis. Moreover, the magnetic layer itself consists of a single sheet of fine, single-domain grains [2]. Usually, the magnetic easy axis of the grains is in the plane of the film; however, a perpendicular easy axis is preferred for the reason of increased data storage capabilities. If the grains are longitudinally oriented, the bulk (or resultant) magnetization of a group of grains forming a *bit-cell* is either oriented to the *left* or *right*. In perpendicular media, this orientation is *up* or *down*. A protective coating made of carbon covers the film stack, which also includes various underlayers below the magnetic film [3]. Nevertheless, due to the fabrication process the magnetic grains have random crystallographic orientation and are of irregular size. If the substrate is crystallographically textured, it influences the direction of the easy axis of the magnetic layer, making reorientation possible along some preferred direction [4].

6.1.1 Quality of Magnetic Tapes

Studies performed on longitudinal magnetic media containing CoCrPt in an SiO_2 matrix have shown that the deposition rate is a key element for the quality of the tape. High deposition rates of, for instance, $10\,\mathrm{nm\,s^{-1}}$ are very desirable; nevertheless, they result in faulty microstructure and harmful heating of the substrate. The SiO_2 matrix was earlier proved to reduce magnetic grain coupling, thereby lowering media noise without any further heating, annealing, or some other processing [5]. In the CoCrPt studies, differences in microstructure grain size and orientation were examined with bright field and dark field transmission electron microscopy (TEM) for varying deposition rates (Figs. 6.1 and 6.2). A slower deposition rate produced a better defined SiO_2 network that minimized coupling between the magnetic grains, results confirmed by a similar experiment [6]. The nonmagnetic phase of SiO_2 separated the grains well, given that boundary segregation of oxides had been reported by other researchers [7, 8] to be an important factor in fabricating well-isolated Co grains. Moreover, oxidation conditions were responsible for the isolation between Co grains and their increased coercivity [9], while media noise performance depended on the oxide type [10].

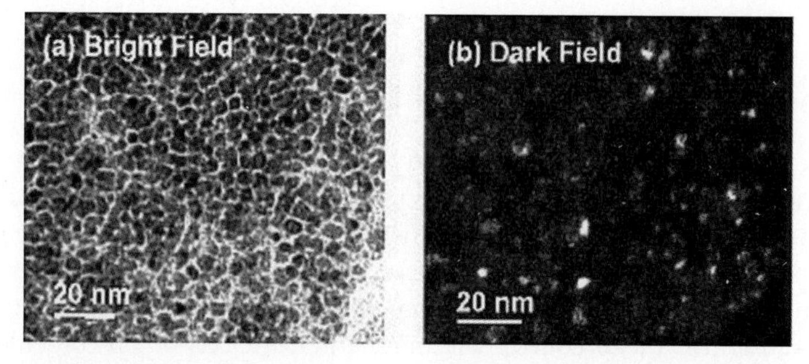

Fig. 6.1. (**a**) Bright field and (**b**) dark field TEM images of magnetic media produced with a deposition rate of $0.28\,\mathrm{nm\,s^{-1}}$(reprinted from [42] (copyright 2006) with permission from the IEEE)

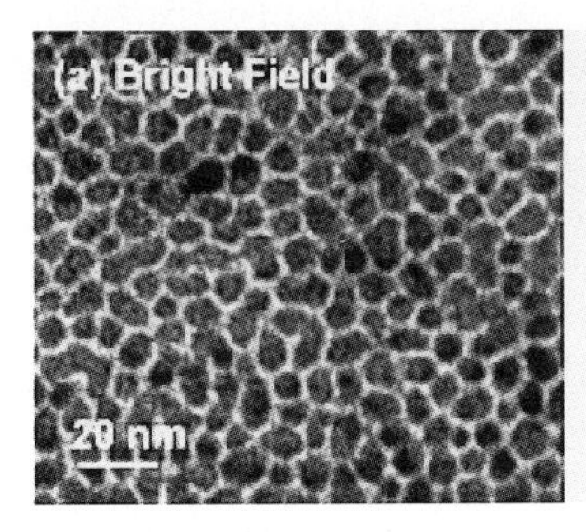

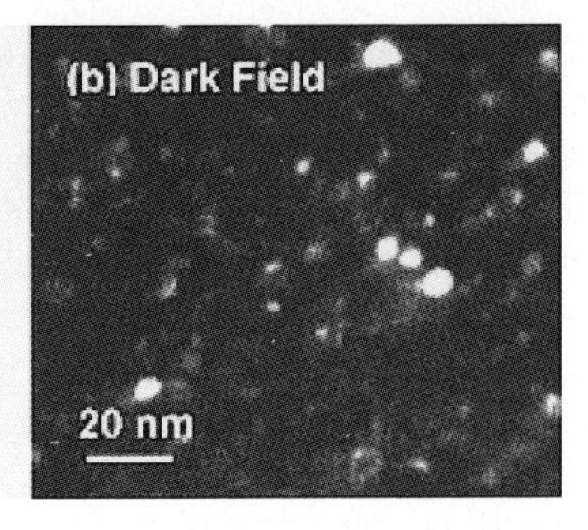

Fig. 6.2. (**a**) Bright field and (**b**) dark field TEM images of magnetic media produced with a deposition rate of 0.05 nm s^{-1} (reprinted from [42] (copyright 2006) with permission from the IEEE)

The thickness of the magnetic layer is expected to result in dissimilarities in the degree of Co grain isolation and grain size, in addition to differences in exchange coupling [11]. Yet, no particular trend was observed for a given thickness as far as crystallographic and magnetic c-axis orientations were concerned [12]. Even so, an optimum signal-to-noise ratio and best possible media writability require a compromise in layer thickness, coercivity, and grain microstructure [13]. Additionally, concessions have to be made in the choice of materials including the intermediate layer [14] between the granular recording film and the soft magnetic underlayer [15].

6.1.2 The Pressure for Higher Capacity Magnetic Tapes

Demands for increased data storage densities have placed more pressure on manufacturers to produce magnetic tape systems with higher capacity. Not long ago, areal recording densities for flexible media had reached 3 Gbits in.$^{-2}$ for metal particulate media [16], and about 11.5 Gbits in.$^{-2}$ for metal evaporated media [17]. To further increase data storage of conventional magnetic tapes, the granular film of the Co-based alloy is reduced in thickness and grain size, allowing larger coercivities [18]. Each grain behaves as a single magnetic particle displaying a certain degree of exchange coupling to its neighbors [19].

6.1.3 Constraints Imposed by Thermal Stability

When the grain size is reduced below ~10 nm, thermal fluctuations become uncontrollable, and a spontaneous reversal of grain magnetization cannot be prevented [20]. The magnetic energy per grain becomes too small to avoid such thermally activated reversals. This is known as the *superparamagnetic effect*, and it leads to the loss of information [21]. If κ_α is the magnetic anisotropy constant, υ is the magnetic switching volume, and kT is the thermal activation energy, then the *thermal stability factor* $\kappa_\alpha\nu/kT$ needs to stay larger than 60 to keep the thermally activated reversal manageable for conventional longitudinal recording media [22].

To maintain a sufficient signal-to-noise ratio for these types of media, the number of grains per bit should not be reduced. Thus, only the grain volume υ can be made smaller. At the same time, κ_α cannot be increased limitless, as the required magnetic field for writing a bit would also need to be enhanced, and with existing head designs only fields that do not exceed 15 kOe can be produced [23]. With these constraints, magnetic heads can currently write only on ~8 nm diameter magnetic grains [24].

Longitudinal granular media have a distribution of small grain diameters that leads to increased thermal activation reversals, predominantly at densities higher than [25] 100 Gbits in.$^{-2}$. For perpendicular recording, the instability of the magnetization due to the onset of superparamagnetism sets the limit to densities of 0.5–1.0 Tbits in.$^{-2}$ [26]. In fact, single domains are expected to form below a certain size due to lesser energy demands for the formation of a domain wall, as compared to those required to create a two-domain state [27]. If the latter were divided by a domain wall, the magnetic island would not be in a stable state [28]. The magnetization would reverse or domains would nucleate when the domain wall would move rapidly [29]. For high-quality recording, the magnetic islands must not reverse magnetization under thermal activation and become superparamagnetic [27].

6.1.4 Forming a Bit

A sufficiently strong external magnetic field can switch the magnetization of individual magnetic grains. Assuming the grains remain stable, if information is recorded on the randomly oriented granular medium, contributions from several hundred grains need to be averaged to form a bit which can be read with sufficiently high signal-to-noise ratio. To improve the latter, it is best to use magnetically textured media where the easy magnetization axes are preferentially oriented along the track [30]. However, if the magnetic tape contains metal particles and they are being reduced in size, it becomes difficult to sustain sufficient coercivity. Therefore, other types of fine particles with large coercivity such as iron nitride [31] and Ba ferrite [32] have been suggested as an alternative.

A magnetic tape containing 21 nm (plate diameter) Ba ferrite particles produced using an advanced precision coating process was recently studied [33] for its improved magnetic recording properties. The volume of the particles was roughly 36% smaller than that of metal particles of 45 nm length. Despite their smaller volume, the coercivity of 160 kA m^{-1} was nearly equal to that of metal particles, and the thermal stability factor was estimated at 65, high enough to ensure stability. Furthermore, the stack of layers was similar to the one used in actual magnetic recording, as it comprised the magnetic layer, an underlayer, a substrate of polyethylene naphthalate base film, and a back coat layer.

A reduction in magnetic layer thickness to 65 nm with less than 10% thickness variation was possible due to the advanced precision coating process. The main result of the study was the enhanced signal-to-noise ratio of the Ba ferrite particles, a value of 23.5 dB compared to 14.5 dB for the metal particulate tape. It should be mentioned that the signal-to-noise ratio for magnetic tapes is roughly proportional to the number of magnetic grains per bit. Therefore, increasing bit areal density without sacrificing signal-to-noise ratio requires reducing the average grain volume in the magnetic medium.

The tapes were evaluated using a metal-in-gap write head with a gap length of 0.2 μm and a track width of 20 μm, as well as a giant magnetoresistive read head with a track width of 1.5 μm and a shield-to-shield length of 0.16 μm. In a typical magnetic recording head, there is an inductive write element, and a magnetoresistive read element. The head is mounted on an air-bearing slider while flying along the data track at nanometer size heights. During the flight along the track, the head stores a sequence of 1 and 0 (i.e., magnetic data) on the magnetic medium by magnetizing into two possible magnetization directions. The information is read back by the read sensor that detects the varying magnetic flux emanating from the magnetized pattern on the track.

6.1.5 Influence of Magnetic Anisotropy

Studies performed in pure and Co/Ti-doped Ba ferrite particles (of hexagonal crystallographic structure) uncovered interesting facts in the role played by the magnetocrystalline and shape anisotropy constants. As expected, the former dominates the latter at all temperatures if the Ba ferrite particles are undoped, giving rise to a uniaxial anisotropy along the *c*-axis. However, the dominance is reversed for the doped particles where the magnetocrystalline anisotropy is weakened by the ionic substitutions, especially at lower temperatures. This happens in spite of the temperature dependence of both anisotropy constants, where the decrease in temperature should have enhanced the magnetocrystalline anisotropy. With the shape anisotropy becoming larger, the overall magnetic anisotropy is no longer uniaxial, presenting multiple preferred directions for the magnetization. To validate comparison with conventional longitudinal particulate recording media, the particles were assembled with aligned *c*-axis in the deposition plane [34].

6.1.6 Choice of Materials

Co/Pd [35] and CoCrPt [36] alloys, particularly if prepatterned, are more likely to respond favorably to the thermal stability problem. This is because the entire volume of the magnetic islands is the effective switching unit, yet only for densities below [37] 10 Tbits in.$^{-2}$. Barium ferrite particles display characteristics that also make them viable candidates for superior magnetic media, because they meet some of the above requirements while offering a chemical stability common to oxide compounds [38]. Lately, areal recording densities of 17.5 Gbits $in.^{-2}$ [39] have been demonstrated for Ba ferrite particulate media by using a giant magnetoresistive head for reading. These results were partly possible because of the advanced fabrication process employed for the magnetic medium. Due to the use of high dispersion technologies for fine Ba ferrite particles, a smooth surface and uniform particle distribution were obtained in an 80-nm-thick magnetic layer. A high signal-to-noise ratio (~19 dB) and increased resolution were observed as well. Further observations confirmed these findings by achieving the same areal recording density of 17.5 Gbits $in.^{-2}$ in Ba ferrite particulate media, and 31 Gbits $in.^{-2}$ in CoCrPt–SiO_2 sputtered media by a different group of researchers [40].

6.2 Bit Patterned Magnetic Media

6.2.1 Bit-Cells

These days, bit patterning the magnetic medium is highly favored for next generation magnetic data storage solutions [41]. Discrete bit media have several distinct advantages over continuous media [42], partly because the discrete bit-cells are more capable of maintaining a thermally stable single domain remanent state [43]. The latter is characterized by a common and well-defined easy axis of constant orientation with respect to the head read/write elements [44]. Additionally, the anisotropy and hence the coercivity agree with the existing write field, while the saturation magnetization can be adjusted to optimize recording [45]. Each bit-cell stores 1 bit of information, resulting in ordered arrays of discrete magnetic entities containing data [46]. Considering that in conventional magnetic media several randomly oriented grains are contained in 1 bit, this contrasts significantly with the fact that a bit of patterned media is formed from one single prepatterned grain (Fig. 6.3). Additionally, superparamagnetic behavior in magnetic grains is avoided in media with patterned bits [47].

By employing a previously defined magnetic entity per bit, larger volume magnetic bit-cells became possible. This reduces the number of magnetic switching volumes per bit to one entity, without the statistical averaging over several grains [48]. As a result, the constraints imposed on the magnetic grains of conventional media are avoided, allowing about 10^{12} prepatterned magnetic bit-cells per surface to be obtained in patterned media. It is estimated that the size and periodicity of the bit-cells are 25 and 35 nm, respectively, assuming an initial recording density of 500 Gbits $in.^{-2}$ [49]. The bit size determines the switching volume, making bit-cells thermally stable, and yet still reversible by the head technology [50]. At the same time, transitions across and down the track are defined by the bit pattern, and no longer by the magnetic field of the head [51].

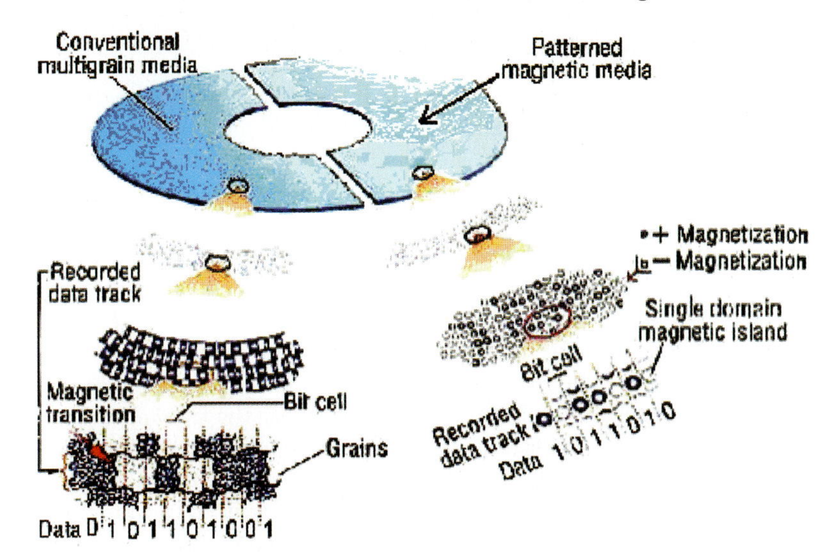

Fig. 6.3. Conventional (*left*) and patterned (*right*) magnetic media. The main difference lies in the fact that for conventional perpendicular media, several randomly oriented grains are contained in 1 bit, whereas the bits of patterned media consist of one single grain that has been prepatterned (reprinted from [49] (copyright 2006) with permission from Solid State Technology and Microlithography World)

6.2.2 Minimizing Errors

The magnetic easy axis of each bit and that of the recording head have the same relative orientation with respect to each other, in contrast to the randomly oriented axes of the grains in the plane of the film in conventional recording media [52]. However, there is a problem with written-in errors that differ from the usual signal-to-noise ratios. The former are due to statistical fluctuations of the magnetic properties and the locations of the individual dots [53]. The size and shape of the patterned magnetic islands as well as their position may vary from one data cell to another, because of lithography or self-assembly errors in nanoparticle size distribution [54]. Therefore, a magnetoresistive playback head of a certain design must be chosen to maximize playback signal resolution and to minimize sensitivity to noise due to these patterning imperfections. Additionally, the playback signal amplitude experiences a linear loss with filling ratio or packing density, which represents a percentage of the bit area filled with magnetic material. It was observed that differential readers exhibited enhanced performance with respect to spatial resolution and sensitivity to medium imperfections due to their ability to compare bits [55].

6.2.3 Some Disadvantages of Patterned Bits

Patterned bits cannot be easily accommodated due to the geometry of the recording media. Circumferential magnetic texturing [56] is required in patterned longitudinal media which is difficult to achieve using polycrystalline films and substrates [57]. A partial solution is to orient the magnetic easy axis normal to the film surface to form a perpendicular recording medium. Magnetic force microscopy (MFM) was initially used to observe the magnetization states of these patterned bits [58], and shortly after it was observed that the tip itself can be used to switch the magnetization of a bit [59]. Nevertheless, an MFM can only be used as a laboratory technique for studying magnetization reversals and domain configurations of patterned media. Otherwise, bits can be reversed inadvertently, as it is only the tip scan height that sets the difference between reading and writing [60]. To better control writing the bits, local heating can be applied, in addition to the tip field [61]. If a current passes between the tip and the island to be written, or an additional field is applied, some of the bit reversal can be prevented [62].

Patterned bits must be organized in a circular configuration, so that the head can remain on a single track of units as it goes over a given area. In addition, servo marks need to be defined on the disks to keep the head on a given path. In conventional recording, these servo marks are written on a plain continuous medium with a recording head, but without predetermined track locations. In patterned media, the locations are predefined which requires the servo marks to be patterned. Because of the circular symmetry of magnetic patterns, Cartesian e-beam lithography developed for mask production in the integrated circuit industry cannot be employed efficiently.

6.2.4 Solutions for Patterning Bits Efficiently

A mechanical rotary stage and a fixed e-beam column ensure patterning of a master mold dot-by-dot while the e-beam resist covered master is rotating. After exposure, the resist is developed and reactive ion etching (RIE) is used to transfer the pattern into the master mold substrate. A disk substrate is coated with liquid nanoimprint resist, and the previously obtained master mold is pressed against this disk substrate. The liquid resist reflows until the master topography is copied onto the resist and ultraviolet (UV) light is applied to cure the resist and create a solid replica. The resist pattern is afterwards transferred into the disk substrate using RIE. Pillars on the replica correspond to holes on the master mold. After resist stripping and disk substrate cleaning, a magnetic medium (Co/Pt [63] or Co/Pd [64] multilayers) is sputter deposited over the patterned disk substrate.

Bit patterned media offer the possibility of increased recording densities, if certain challenges are overcome. For instance, imperfections in bit spacings, size, thickness, and shape, as well as fluctuations in saturation magnetization result in a signal-to-noise ratio similar to that of conventional recording media [65]. In recent years, *perpendicular discrete bit media* have received progressively more attention as they display a higher density potential due to larger head write fields, and enhanced thermal stability of single domain islands. The latter have the advantage of eliminating the jitter noise of conventional media, which usually arises due to grain irregularities in the transitions. Nevertheless, the major difference between discrete bit and conventional media is the necessity of synchronizing the head position with the predefined island locations [66].

As mentioned, the main source of noise is likely to originate in bit size and bit pitch variations [67, 68]. Therefore, the write field has to be well placed to write on the desired bit and not on adjacent bits [46]. Cleanliness and smoothness are also issues, as they can interfere with the ability of the head to fly at the required sub-10 nm fly height [69]. Additionally, the thickness of the medium needs to be of the order of the bit spacing to prevent the head from inadvertently writing neighboring areas with the fringes of the head field gradient. Stray fields from adjacent bits scale with saturation magnetization, and so does the contribution of the shape anisotropy to the total uniaxial anisotropy [70].

6.2.5 Materials for Bit Patterned Magnetic Media

The most likely candidate materials for perpendicular discrete bit media are those with interfacial anisotropy such as Co/Pt [71] or Co/Pd [72] multilayers, as well as CoPtCr alloys [73] grown with the *c*-axis normal to the surface. High exchange coupling between the individual entities is needed if single domain areas are to be achieved, in contrast to the low exchange that is important in conventional magnetic recording [74]. Patterned bit media do not demand a magnetically soft underlayer as previously used for continuous films to channel flux between a small write pole and a large flux return pole. This underlayer would add unneeded complexity to the patterning of the bits, however, it may become again necessary if densities in excess of 1 Tbit in.^{-2} are envisioned.

FePt is a material that instead of being magnetic in the chemically ordered state, it is magnetic in the disordered state [75]. However, it has the advantage that nanoparticles as small as 4 nm diameter have sufficient magnetocrystalline anisotropy to remain magnetically stable at these small volumes, avoiding superparamagnetism [76]. Lately in a series of experiments, FePt nanoparticles were fabricated through solution chemistry and seized precipitation [77], the process resulting in a suspension of surfactant coated particles [78]. A substrate dipped into the suspension became covered with FePt nanoparticles that formed self-assembled nanolayers [79].

The FePt nanoparticles were oriented at random [80] and were in a disordered fcc crystallographic phase in which they were magnetic [81]. In some cases, annealing at temperatures higher than 600°C was required to induce a phase transformation [82] with the nanoparticles displaying a shared magnetic easy axis [83]. Nevertheless, this latter step can lead to the undesired consequence of particle agglomeration [84, 85]. Alternatively, this effect can potentially be eliminated by surfactant chemistry optimization [86], or by hardening the organic matrix with ion beam irradiation [87]. In any case, it is not desirable that some nanoparticles should remain in a disordered fcc phase while others attain some degree of ordering. Therefore, it is necessary to prevent a large scale spreading of anisotropy in the medium [88].

Some studies suggest that ordering depends on particle size [89]. At the same time, there is a chance that alloying with a metallic element may also lead to ordering [90]. In a recent experiment involving chemically ordered antiferromagnetic $FePt_3$ nanoparticles, ion beam irradiation with 700 keV N^+ ions at a dose of 1×10^6 ions cm^{-2} destroyed the chemical order and induced ferromagnetism [91]. Concurrently, it was also observed that this method preserved film crystallinity, therefore showing potential to be used for patterning magnetic films.

6.2.6 Maintaining Competitiveness

Any new nanopatterning methods that are expected to be competitive in the recording media market need to satisfy the four criteria: inexpensive fabrication, capability to produce circular symmetry patterns, have the potential of sub-50 nm resolution, and result in a clean surface for the head to fly over. Lithography processes may meet these requirements; however, they all have their limitations.

Optical lithography has a resolution given by $\lambda/(2\mathrm{NA})$, where λ is the light wavelength used and NA is the system numerical aperture. Various resolution enhancement techniques can be employed, so that a wavelength of 193 nm can achieve 90 nm line widths, with possible reductions to 45 nm using immersion lithography, nevertheless still not sufficient for patterning the recording medium. Only extreme UV light or X-rays can achieve the required resolution, however, these techniques are not commercially available due to their prohibitive costs.

6.2.7 Going Nano and Beyond

Interference lithography has been proposed as an alternative for disk patterning [92]. Among its accomplishments are magnetic Co nanodots [93], and Co or Ni nanopillars [94] that seem to be studied often [95]. Interference lithography is maskless, as the patterning is done by interference fringes produced by two optical beams giving rise to a pattern period $p = \lambda/2\ \sin\theta$, where θ is the half-angle between the two beams. Patterned dots have been obtained through successive 90° rotations of the sample between exposures, resulting in patterned areas whose size depends on the power of the laser beam. Consequently, centimeter scale arrays of magnetic nanodots [96] have been successfully fabricated using interference lithography. In addition, the technique can be used for producing a master for nanoimprinting [97], or for local annealing of previously deposited films [98].

Due to the limitations imposed by the $\lambda/2$ minimum period even at short wavelengths, the sub-50 nm resolution required for discrete bit recording is difficult to achieve by interference lithography. By using a series of diffraction gratings for producing interference minima and maxima, achromatic interference is obtained that can overcome the wavelength limit of monochromatic interference. Magnetic nanodots have been patterned this way while the fringe pitch was 100 nm [99]. Another advantage offered by diffraction gratings is that they can create circular beams not otherwise accessible due to the linear fringes of conventional interference lithography. Thus, patterns with circular symmetry have been demonstrated [100].

Magnetic nanostructures of sub-100 nm size have been successfully obtained using ion beam bombardment [101] or electron beam (e-beam) lithography [102], although both are expensive and slow writing processes. The resolution is not determined only by the e-beam diameter, but also by resist properties responsible for proximity influences of the overlapping low level exposures. These occur due to electron scattering in the resist and substrate [103]. In spite of its shortcomings, e-beam lithography should not be dismissed as a viable technique for the fabrication of masks used in master/replication methods.

X-ray lithography, in particular that using soft rays (1–10 nm wavelength), also uses a mask fabricated using e-beam lithography, and since X-ray lithography is a parallel process [104], an area as large as the mask can be patterned in a single exposure. Nanoscale features such as 88 nm permalloy dots [105] or 200 nm Co dots [106] have been patterned. Nevertheless, X-ray lithography requires a high intensity X-ray synchrotron.

Cost and accessibility remain issues with lithographic processes currently employed in patterned bit media production. While continuous media disks cost only a few dollars, nanopatterning can make disk fabrication economically unacceptable. In addition, patterned features on a nanoscale are challenging to create even at higher costs. A bit areal density of 300 Gbits in.$^{-2}$ necessitates a 30 nm bit-cell, and under the same prerequisite 1,500 Gbits in.$^{-2}$ will require a 14 nm cell [144]. Thermally assisted recording [107] or lithography technologies for such feature sizes are not probable to be accessible or cost effective soon enough for the hard drive industry to maintain the current compound annual rate of areal density growth.

6.3 Self-assembly and Magnetic Media

6.3.1 Alumina Templates

Self-assembled templates formed through a natural process such as self-ordering offer an appealing alternative to expensive lithography methods not only because of cost, but also because of their versatility. Among the favored materials is anodic alumina due to its ease of production where only variables such as voltage [108], current, [109] and pH of the employed acid [110] control

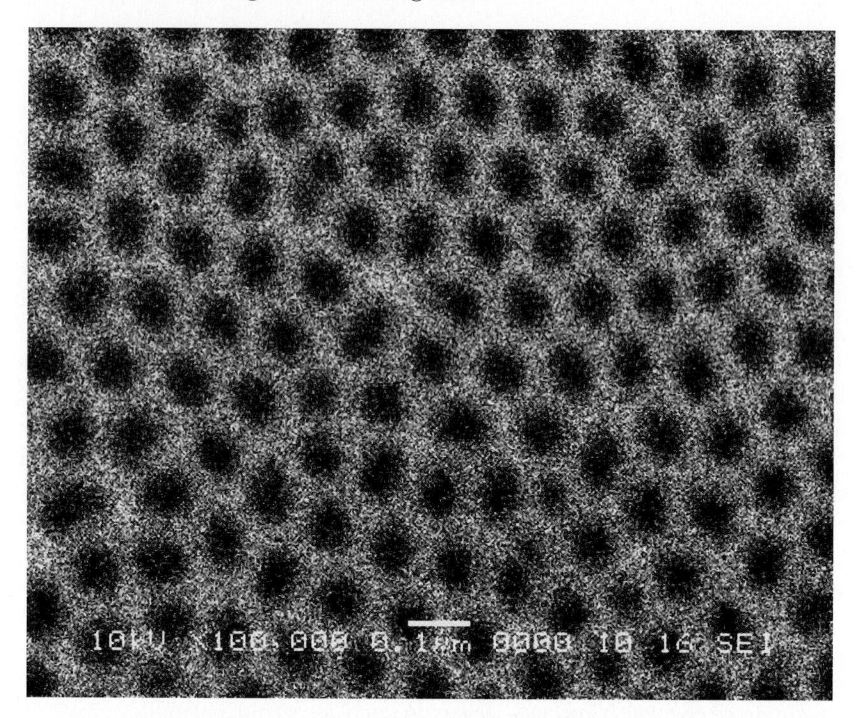

Fig. 6.4. Scanning electron microscopy (SEM) image of an alumina mask/membrane self-assembled through a simple anodization process. The alumina membrane displays throughout its thickness a hexagonal array of nanopores similar to a honeycomb structure (C.-G. Stefanita and S. Bandyopadhyay, unpublished)

the fabrication process. Regular hexagonal arrays of nanometer sized pores (most commonly 10–100 nm [111]) can be formed on the surface of aluminum (Fig. 6.4).

While the pores reach through the whole layer of anodized aluminum, there is a thin Al_2O_3 (alumina) barrier at the bottom of the pores forming an impediment to electrodeposition [112] or evaporation of any material that is supposed to fill the pores [113]. The barrier can be conveniently removed through etching under a reverse dc current and low voltage in a diluted acidic solution [114]. Furthermore, the whole alumina layer, now a template containing the regularly arranged pores [115], can be removed and placed on a different substrate. Prior to transfer, the alumina layer can be thinned (Fig. 6.5) to allow e-beam evaporation of a material of choice. Figure 6.6 shows an example of e-beam evaporation of a metal such as Zn, forming a quasiregular array of nanodots.

The versatility of alumina templates is also demonstrated in Fig. 6.7 where RIE has been used to replicate the hexagonal pattern on a GaN substrate. If the pores are filled by ac or dc electrodeposition, depending on pore diameter

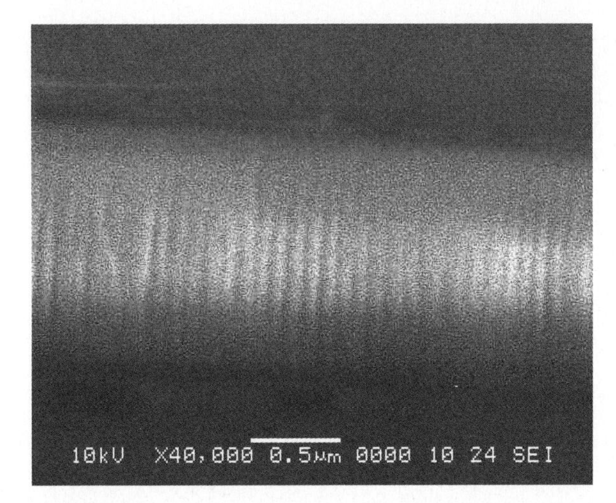

Fig. 6.5. SEM image of a cross section of a thinned alumina template displaying regularly spaced nanopores reaching through the thickness of the template (C.-G. Stefanita, F. Yun, and S. Bandyopadhyay, unpublished)

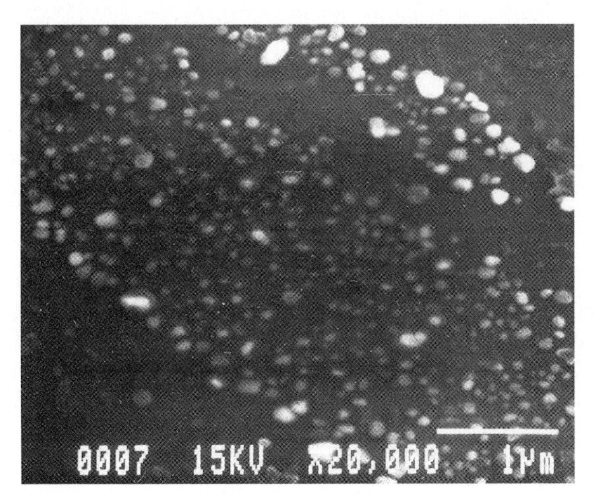

Fig. 6.6. SEM image of quasiordered Zn nanodots obtained by e-beam evaporation through a self-assembled alumina mask. If an electron beam of a heated material strikes the nanopores of an alumina mask, it will deposit minute quantities of that particular material, if the mask is thin enough to allow the beam to penetrate through the pores. The mask is removed after the evaporation and an array of nanodots is obtained on the substrate on which the mask has been placed (C.-G. Stefanita and S. Bandyopadhyay, unpublished)

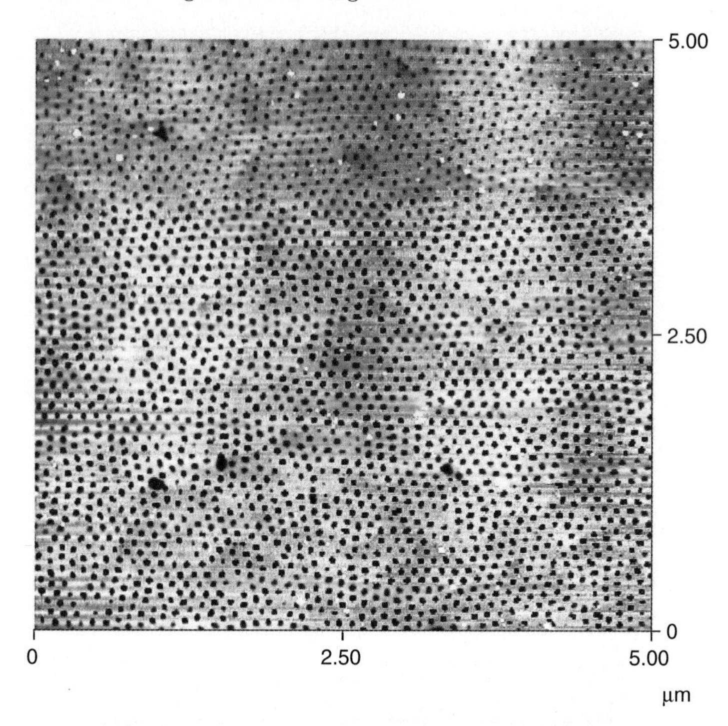

Fig. 6.7. The regular hexagonal pattern of the alumina mask can be etched onto a substrate, as seen in this atomic force microscopy (AFM) image of a duplicated honeycomb nanopattern on GaN. In general, substrates can be etched on a nanoscale using reactive ion etching and an alumina mask that has been appropriately thinned (C.-G. Stefanita and S. Bandyopadhyay, unpublished)

and template thickness, nanowires of a wide range of diameter and height can be obtained (Fig. 6.8). On the other hand, etching the alumina template in a dilute acidic solution for a well-defined time, just before the template dissolves completely, leaves a collection of hollow alumina nanorods behind (Fig. 6.9).

6.3.2 Guided Self-assembly as a Solution to Long-Range Ordering

Given their versatility, self-assembled templates may be the key to obtaining storage densities in excess of 1 Tbit in.$^{-2}$ that require sub-14 nm resolution, as well as new shapes and profiles. When used as a mask, self-assembled templates [116] can be used for etching or depositing just like an exposed resist. The ordered regions can be over 100 μm in size, similar to grains in conventional magnetic recording. Furthermore, the length scales of self-ordering can be increased by templating the surface before further processing using ion beam exposure [117] or interference lithography [118]. The idea of combining lithographic processes with natural templates in what is known as *guided*

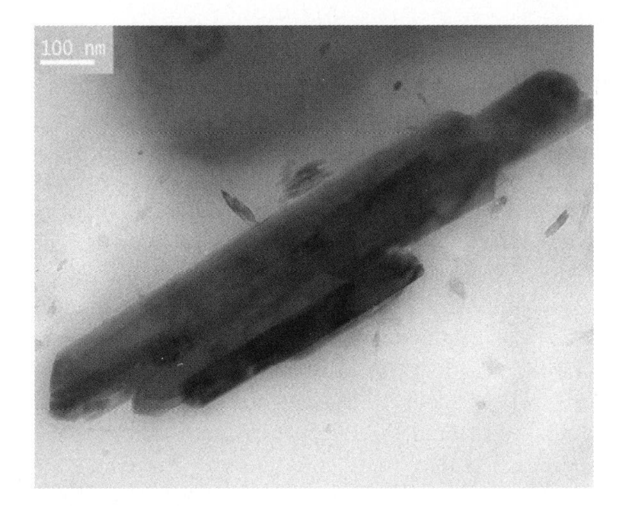

Fig. 6.8. The alumina membrane can be used to selectively electrodeposit under ac or dc conditions metals, semiconductors or organic materials into the nanopores, as seen in this transmission electron microscopy (TEM) image of ZnO nanowires. Arrays of nanowires can be fabricated by electrodeposition in templates that are subsequently removed by etching (C.-G. Stefanita and S. Bandyopadhyay, unpublished)

Fig. 6.9. SEM image of hollow alumina nanorods obtained after etching an alumina template just for enough time to recover the pores without completing removing their surroundings (C.-G. Stefanita and S. Bandyopadhyay, unpublished)

self-assembly is one possible solution to the problem of coherent long-range ordering. There are two categories of techniques that qualify as guided self-assembly: *chemical* where the surface chemistry is selectively modified to allow formation of nanostructures in some areas [119–121], and *topographical* where the surface geometry is altered [122].

6.3.3 Chemically vs. Topographically Guided Self-assembly

In *chemically* guided self-assembly, the surface of the substrate is selectively functionalized (using, e.g., polydimethylsiloxane) so that self-assembled monolayers (SAM) develop, aiding the formation of patterned structures [123]. Conversely, a *topographically* modified substrate requires employment of a combination of masks, stamps, and etching to achieve the desired prepatterning. As an example of topographically guided self-assembly, a sputtered perpendicular $Co_{74}Cr_6Pt_{20}$ 40 nm thick medium was resist-coated and imprinted with an Ni stamp. Additionally, a bit pattern was created in the resist grooves using a polystyrene PMMA block copolymer. After annealing the PMMA, the spheres that formed were selectively removed using an oxygen plasma treatment. Consequently, holes formed in the polystyrene matrix were filled with spin-on-glass. The matrix was subsequently removed by ion milling, and so was the excess magnetic medium, leaving tracks of magnetic nanodots protected by the spin-on-glass [124]. Packing densities of 450 Gbits in.$^{-2}$ have been reported for magnetic nanodots fabricated in self-assembled templates [125]. In an opposite type of experiment, thin films were deposited on top of alumina masks, giving rise to regular arrays of holes in the film, termed "antidots" [126].

6.3.4 Biological Self-assembled Templates

Biological self-assembled templates have also been considered as possible alternatives for data storage media [127]. In particular, the protein *ferritin* offers promises for magnetic particle growth due to its capability of self-assembling into a hollow sphere with an ~8 nm cavity. Inside the cavity, FePt or CoPt nanoparticles can be grown in an *hcp* structure with an arrangement period influenced only by the ferritin dimensions and not the particle within the ferritin cavity. However, the easy axes of the particles are not easy to orient, and the packing fraction is less than that for conventional recording media [128].

6.3.5 The Versatility of Block Copolymers

As already mentioned, block copolymers are versatile compounds used frequently to generate template surfaces through guided self-assembly. They consist of a polymer chain with two distinct monomers bound together [129]. Because of their flexibility, block copolymers can be used as masks to prepare the substrate prior to the deposition of a magnetic thin film. They are usually spun onto a surface, previously prepatterned by interference lithography or RIE. The copolymer migrates to the grooves, leaving the top of the patterned lines free of polymer material. Through subsequent treatment, the block copolymer phase separates producing periodicities of 200–1,500 nm [130], or even finer such as 10–200 nm [131].

The obtained geometry of the periodicities (whether spheres or lamellae) depends on the ratio of the two monomers and the surface energy of the substrate [132]. These periodicities are then utilized as templates for the growth of magnetic nanodots [133], however, sometimes the nanodots themselves are employed as an etch mask, for instance, in the case of a previously deposited Co film with W/SiO_x caps [134]. In a similar copolymer experiment, it is the gap between the 200 nm hexagonally packed polystyrene spheres that is used for the deposition of Co/Pt multilayers forming a reverse structure [135]. It was also observed that the curvature of the sides of polystyrene spheres can be exploited to alter the properties of Co/Pd films [136] deposited in the form of multilayers, and thereby displaying a perpendicular anisotropy, highly desired in magnetic recording [137].

6.3.6 Inorganic Templates May Still Be Competitive

As a substitute for polymer templates, inorganic particles are still used frequently, being capable of forming close-packed arrays on certain types of functionalized surfaces (thereby chemically guided for self-assembly). For example, amorphous SiO_2 was turned into spheres on which Co was deposited and oxidized on the walls of the spheres, creating isolated Co nanodots [138]. However, the geometry of the recording medium, in particular, the rotating disk layout imposes the added constraint of tracks with circular symmetry [139] and nanostructure ordering over large areas [140].

6.4 Present Alternatives for Discrete Media Production

No matter what fabrication technique is utilized or what type of substrate is chosen, there are still challenges that need to be overcome before self-assembled templates can be reliably employed in discrete bit media production. In the meantime, the magnetic recording community is using *master* fabrication and inexpensive *replication* to manufacture CD-ROMs. The master is produced using a laser writer, a rather costly process. Even so, the mass manufacture of CD-ROMs by injection molding from the master is economical and reliable at present.

6.4.1 Patterning with Stampers and Masks

At least for now, nanoimprinting offers a viable alternative to expensive fabrication by using a rigid or flexible mold/stamper [141] to emboss a relief pattern into a thermosetting or photocuring polymer resist layer [142]. After the resist is patterned with the stamper, the former is used as an etch mask to further transmit the pattern onto the underlying layer [143]. The patterned resist layer acting as an etch mask can be a PMMA layer that was heated above its glass transition temperature and onto which the stamper was pressed. Commonly employed materials for the stamper are Si or SiO_2 if the stamper is rigid [144]. Self-assembly can be utilized to pattern the mold or stamper, however, e-beam lithography is routinely employed. Additionally, the primary stamper can be duplicated using nanoimprinting, and the ensuing replica can be used to produce more molds/stampers, such that the duplicated stampers become more economically viable [145].

The patterned layer has to be thick enough to shield the unexposed areas underneath, at least as thick as the necessary etch depth. Trilayers and complex processes which utilize windows are sometimes the answer for better feature depth control [146]. The substrate below the SiO_2 film to be patterned by RIE is typically a silicon wafer with a thin metal layer on top. The role of the metal layer is to ensure adhesion of the SiO_2 film above it.

Ni or Co pillars with perpendicular anisotropy are usually electroplated in the patterned openings, after which the surface is chemically or mechanically polished to achieve a certain smoothness [147]. In an alternative approach, the pillars can be formed by lift-off after depositing a Ni or Co film onto a nanoimprinted PMMA mask [148].

6.4.2 Cleanliness Concerns

Patterned media require much less severe lithography constraints than silicon processing where exact alignments, transfers, and repeat exposures add several important steps to fabrication. There is, however, the issue of exterior smoothness in disk drive applications, as the recording head is supported by an air bearing of sub-10 nm size while flying above the surface at velocities greater than $40\,\mathrm{m\,s^{-1}}$. For this reason, the surface needs to be free of contamination such as resist residue.

The cleanliness and smoothness of the surface are key elements in hard disk drive operation; therefore rigorous requirements are enforced during the fabrication process of patterned media [149]. Even the smallest resist contamination can cause a serious collapse in head performance due to the breakdown in the aerodynamic air bearing above the disk surface. The sub-10 nm head disk spacing supporting the head leaves very little room for error while the magnetic recording head is flying above the surface of the disk [150].

Only about 0.2–0.4 nm rms surface roughness are allowed, posing a challenge to fabrication techniques that necessitate a thorough cleaning process without damaging the magnetic film [151]. The latter is usually prone to corrosion, and quite often too soft to support a flying head. A hard protective layer such as carbon is typically employed, with the additional requirement of edge protection when patterned elements are involved.

6.4.3 Obtaining High Aspect Ratios

In discrete bit and equally in discrete track media, it seems that a good compromise for maintaining such small head disk distances is to pattern the substrate rather than the magnetic film [152]. By creating deep islands and channels in the disk, the side walls of these etched depressions serve to decouple neighbors from each other [153]. Thereby, the magnetic flux emanating from the channels is greatly reduced when it reaches the read element, without interfering with that of the islands [154]. The disk can be easily cleaned after patterning without disturbing the magnetic film, as it does not experience any of the etching processes. However, some protective coating is recommended.

It is difficult to use RIE on magnetic layers, in particular when nanoimprinting is employed for resist patterning, as the high aspect ratio necessary for obtaining nanopillars is hard to achieve. Even if this impediment is overcome, releasing the mask is not trivial considering that the adhesion force is high in the increased pillar contact areas. One possible solution is to put an SiO_2 layer onto a plating base, etch it and subsequently deposit a magnetic film onto the imprinted layer, followed by lift-off to expose the nanodots [155]. Or, as mentioned before, etch the substrate thereby patterning it and then deposit a magnetic film while the side walls of the trenches provide the necessary domain isolation [156]. Imprint patterning and etching the substrate have produced 30 nm Co/Pt nanodots with perpendicular anisotropy [157], and discrete tracks in NiP-coated Al disks on which CoPtCr films were deposited [158].

6.4.4 Types of Nanopatterning Processes

Above procedures can be regarded as *additive* processes for nanopatterning, as the nanopattern is first created and then the magnetic film is deposited. Generally, the resist or some other layer is patterned into an array of holes, after which the latter are filled with a magnetic material by using evaporation or sputtering. If using resist, it needs to be removed at the end through lift-off where the patterned resist is dissolved in a solvent. The deposited film on top of the resist is removed along with the resist.

In contrast to additive nanopatterning, *subtractive* processes involve magnetic films contained underneath layers meant to serve as masks, and any etching that is done to these layers affects the magnetic film as well. The patterned layer turned etch mask ensures that excess magnetic film is removed by ion milling, sputter etching, or wet chemical etching, considering that magnetic materials are difficult to etch using RIE. Ion milling is not a greatly selective process, and while different materials have similar etch rates, thick polymer masks or hard masks made of metallic films have to be employed. For these processes etch masks obtained by self-assembly can be used as an economical alternative to more expensive masks [159].

6.4.5 Emerging Fabrication Techniques

In spite of the difficulty of directly patterning the magnetic structures, several *prototype techniques* have been utilized, among them *focused ion beam* (FIB) using Ga^+ ions [160]. Patterned servo marks [161] or 100 nm patterned bits [162] have been etched in conventional longitudinal recording media using FIB. Irradiation with light ions such as He^+ [163] or Ar^+ [164] has been employed for spatially modulating the anisotropy of several magnetic systems. Initially ordered films of FePt can be disordered to some degree if ion irradiation is combined with thermal annealing, resulting in areas with higher anisotropy [165]. Additionally, it was noticed that Ga^+ implantation in CoPtCr films resulted in single domain remanent states if the islands were smaller than 70 nm in diameter [166], or that the interfacial anisotropy of Co/Pt multilayers is reduced by FIB [167]. Nevertheless looking on the plus side, the disk surface was sufficiently clean to fly a recording head above it and permit comparison of readback waveforms from patterned and conventionally written bits.

Contact and proximity masks have been used to pattern exposures of light ions. For He^+ ions of 30 keV resist masks of 600 nm are necessary, whereas tungsten masks need to only be 200 nm thick [168]. A viable alternative may be offered by selectively blocking ion exposures via contact resist or masks placed directly on the films, such as membrane masks made of etched silicon [169]. A few hundred nanometer thick Si membranes are sufficient for stopping light ions of tens of kiloelectronvolts [170].

Fabricating membranes for 2.5 in. diameter disks with the sub-50 nm size requirements for discrete bit recording are quite demanding. *Projection ion beam lithography* may help to reduce the feature size in the masks used in FIB, as the diffraction limit of ions does not limit the achievable resolution. With ion wavelengths of the order of 10^{-5} nm, the diffraction limit is only a few nm.

An optimistic resolution limit for SiO_2 masks was reported to be 14 nm [171], which is higher than the *magnetic resolution.* The latter is limited by the exchange coupling in the magnetic film, nonetheless highly desirable if single domains are to be formed. Magnetic domain walls may propagate from low anisotropy to high anisotropy regions, resulting in poorly defined boundaries and less stability [172]. Nevertheless, it is the dense magnetic feature concern, and the fact that projection systems would require masks ten times larger than the disk diameter if patterning is to be done in one exposure, that have so far limited the use of these techniques [173].

Even more unusual for nanopatterning magnetic films is using a conventional AFM tip coated with a barium ferrite solution. The technique is known as *dip-pen nanolithography*, as the AFM tip is brought into contact with a suitably prepared substrate, resulting in sub-100 nm size features [174].

Although these laboratory tools involving direct writing or etching using focused ions, or some other resourceful idea have been successful at fabricating structures over small areas [175], they lack the high throughput necessary for the industrial fabrication of patterned media.

6.4.6 Discrete Track Media

Some proposals advance the idea that discrete track media can be regarded as an efficient way to increase areal densities. For instance, using an e-beam defined resist mask a Co containing magnetic film can be etched, thereby patterning the surface of the film into tracks [176]. Thus, magnetic properties of discrete track media are comparable, yet only up to a point to those of continuous thin films. What they have in common is the similar manner of writing transitions into the tracks. The transition width between adjacent bits, and consequently the linear recording density depend on the size and uniformity of the grains. However, write fringing, erase band, and track-edge noise limit the feasible track density [177].

If the tracks are physically defined, a greater signal-to-noise ratio may be achievable, although writing resolution at the track edges remains a critical challenge. Even if magnetic isolation of grains improves the write resolution at the track edges, with narrow track recording of ~100 nm or less expected for densities of 1.0 Tbits in.$^{-2}$ and beyond [178, 179], track edge irregularities are likely to contribute to the track-edge noise. Therefore, at these track widths e-beam lithography is unavoidable, making the additional processing not cost effective.

Maybe embossing or nanoimprinting shows more promise, as they involve patterning the substrate prior to thin film deposition [180]. Grooves can be first etched into a layer covering the disk, and then a magnetic film can be deposited on top of the patterned film. If the grooves are deep enough, there is reduced likelihood of a readback amplitude.

With fabrication issues remaining the main problem, it is up to techniques that have industrial suitability to demonstrate that discrete track patterning has potential.

6.4.7 Identifying Track Locations

The role of a magnetic servo pattern is to define the track positions in order for the magnetic head to repeatably write and read the data bits. During the writing process, the head is flown over the entire disk surface while the servo marks are being recorded, track by track. The recorded servo marks identify the track locations, while the write field from the recording head delineates the bit length along the track and the bit width across the track. This process is both slow and expensive, in contrast to the stamping or injection molding of optical media such as CDs and DVDs.

There can be a problem of misregistration of the head to the track location after each revolution, due to mechanical vibrations and imperfect head positioning. By patterning the tracks and hence physically making out the track locations on the disk, the errors in head tracing can be reduced. Regions between tracks should be nonmagnetic or at a lower level than the head, to prevent the read element from sensing any magnetic response from them.

Largely, the read element can be manufactured of comparable width with the write element without concern for side reading, thereby improving fabrication yield and lessening requirements on head tolerances. The track width is still lithographically defined; however, the down track bit transitions are determined by the field gradient from the write head, making material requirements for discrete track media similar to those of continuous media [181]. Granular CoPtCr-based films [182] deposited onto either prepatterned glass or NiP substrates make it common place to employ conventional magnetic recording materials for discrete track media [183].

6.4.8 Parallel Writing of Data

Initially, preembossed rigid magnetic disks (PERMs) were used for parallel magnetic data recording, where the data were recorded on the lands of a land/groove structure, while the embossed pattern contained track following marks [184]. The whole idea was to write equal information on several disks simultaneously, and that included formatting the media, such as inscribing sector headers or position marks. The PERM disks were manufactured by embossing and etching a land/groove pattern into a glass disk. A magnetoresistive element detected an error signal representative of the track following on these large grooves. Feed forward compensation, achieved by mounting a prepatterned disk onto a spindle, was used for track eccentricity.

6.4.9 Magnetic Lithography for Mass Data Replication

To improve the patterning of servo information or for more efficient mass replication of content onto magnetic disk media, several fabrication techniques have been developed, among them *magnetic lithography* [185]. Using a master or a mask, the magnetic pattern is transferred onto a substrate referred to as a *slave*. The latter is first magnetized uniformly in one direction by an external magnetic field, after which the mask is placed in closed contact with the disk, and the direction of the external magnetic field is reversed. During this process, the magnetic mask selectively shields the slave disk from the external field while the magnetic pattern is transferred onto the disk [186].

Magnetic patterns can be transferred via rigid masks with air channel and vacuum [187], or rigid Si masks with flexible disks such as metal tape [188]. Alternatively, a magnetic pattern can also be transferred by creating a master while etching a substrate and coating the surface with a soft magnetic film, covering the grooves and lands. The gap where the substrate is etched prevents magnetic shielding from the grooves, and the pattern can be transferred [189]. Masters employed in magnetic lithography can be used over one million times to print disks [190]. Even perpendicular recording media with coercivities as high as 10 kOe have benefited from the printing technique [191]. Nevertheless, transferring patterns by magnetic lithography is not likely to reach the tens of nanometers resolution required for high density recording.

6.4.10 Magnetic Disk Drives vs. Semiconductor Processing

DRAM or semiconductor memory was sold for $100 per GB in 2005, whereas magnetic disk drives were only $1 per GB, in spite of their mechanical complexity [192]. This discrepancy is due to the 10^{10} features usually necessary on a semiconductor memory wafer, as compared to only 10^5 typically encountered on 5-in.-diameter wafers of magnetic heads (~40,000 heads). Magnetic recording heads are fabricated with fewer processing steps than semiconductor-based memories. It is only the track width that needs to be lithographically defined at the minimum size (~120 nm, half the track pitch in 40 Gbits in.$^{-2}$ disks), in contrast to the down track spacings between transitions that are defined by the thin film deposition process. In this case, a single head and disk may be able to address 30 GB of data [193].

6.4.11 Head Performance

Head performance is typically evaluated as a function of an external variable such as pressure that translates into a determination of spacing. It was argued that perpendicular recording is less sensitive to head spacing than conventional longitudinal recording due to the soft underlayer [194]. However, recent experiments have demonstrated that the two recording techniques have similar sensitivities. To prove this, the projects involved using drives with the same air-bearing design that allowed the drives to respond with a similar change in spacing when the ambient pressure varied.

Specifically, the longitudinal write heads incorporated a dual layer eight turn coil, a 15 μm yoke, and a physical write width of 215 nm. Conversely, the perpendicular write heads contained a single pole, while the physical write width was 150 nm. In both types of drives, the read heads were giant magnetoresistive spin valve sensors with a Co–Fe–Ni–Fe free layer and a physical read width of 115 nm. The drives were placed in a pressure chamber where variations either above or below atmospheric pressure raised or lowered the nominal flying height. The performance of the drives was assessed based on the raw uncorrected bit error rate, or soft error rate [195].

The magnetic field dynamics and spatial distribution under the recording pole control the perpendicular recording process. Nonuniform pole footprints accompanied by cross- and down-track degradation of media saturation and overwrite were observed occasionally [196]. Consequently, measurement techniques were developed to assess the stability of perpendicular recording heads. Heads with well-saturated uniform media saturation under the recording pole result in a symmetrical cross-track overwrite profile, whereas heads with nonuniform footprints generate asymmetrical cross-track overwrite distribution [197]. The head keeper spacing is the head air-bearing surface to soft underlayer surface, and is one of the critical parameters that affect the writer field strength [198] and the playback process [199]. A nondestructive method for determining the head keeper spacing from the playback waveform of a specific recording pattern was developed for this purpose [200].

6.4.12 Spin Valves and Giant Magnetoresistive Heads

Given their low intrinsic impedances, all-metallic giant magnetoresistive heads in a current-perpendicular-to-plane geometry are appealing for future high density recording sensors. Nevertheless, these impedances are increasing when the cross sectional area of the sensor is reduced, a necessary aspect in keeping up with rising recording densities. To solve the problem, magnetic heads have been redesigned by employing a *dual spin valve* sensor stack with thin IrMn antiferromagnetic top and bottom pinned layers, as shown in a recent study by Hitachi GST researchers [201].

In a different study, *corrosion resistant* spin valve films made of Ta/NiFe/CoNiFe/CuAu/CoNiFe/PtMn/Ta built into conventional shield-type giant magnetoresistive heads, and tested using a helical-scan tape drive showed a preservation of magnetic properties over 20 days. This occurred despite being exposed to corrosive gas [202], an estimated equivalent of environmental pollution a personal computer experiences over 10 years [203].

Further obstacles need to be overcome in the design of giant magnetoresistive heads, among them an increased signal-to-noise ratio and better resolution, if these heads are to be employed as future recording sensors.

6.4.13 Looking Back and into the Future

Since 1956 when IBM launched in the market the first magnetic hard disk drive capable of storing about 4.4 MB, the storage density (measured in bits in.$^{-2}$) has increased more than 2×10^7 times [204]. This has been achieved primarily by scaling the components of the disk drive to smaller dimensions so that the head moves closer to the disk, enhancing the write and read resolution [205].

Improvements in head design have increased their performance [206], which means that high coercivity media can be written at high data rates. Yet, bit areal density growth has slowed down from 100% per year experienced over a decade or so to 40% per year in recent years, due to remaining challenges encountered in recording head design, as well as magnetic media fabrication [49].

There seems to be an agreement in the magnetic recording industry that giving up the requirement for multiple grains per bit may settle the thermal stability issue. This leads to the necessity of adopting alternative fabrication and magnetic recording technologies to achieve even higher densities. While more traditional magnetic recording approaches start to encounter physical limitations set by the thermal stability of the recorded bits, *nanofabrication* is offering new solutions for increased data densities [207].

Self-assembled arrays of high anisotropy magnetic nanoparticles may offer a replacement for thin film magnetic media, especially if they can be produced extremely small in size [208]. It seems that new doors for higher storage densities and novel disk drives not attainable by scaling current technologies may open if lithography or self-assembly are used for disk patterning [209]. It is currently envisioned that magnetic nanostructures will be the basic units for future patterned media.

References

1. J.G. Zhu, IEEE Trans. Magn. **28**, 3267 (1992)
2. K.M. Tako, T. Schrefl, M.A. Wongsam, R.W. Chantrell, J. Appl. Phys. **81**, 4082 (1997)
3. J.O. Oti, IEEE Trans. Magn. **29**, 1265 (1993)
4. J.J. Miles, B.K. Middleton, IEEE Trans. Magn. **26**, 2137 (1990)
5. T. Oikawa, M. Nakamura, H. Uwazumi, T. Shimats, H. Muraoka, Y. Nakamura, IEEE Trans. Magn. **38**(5), 1976 (2002)
6. L. Wang, H.-S. Lee, Y. Qin, J.A. Bain, D.E. Laughlin, IEEE Trans. Magn. **42**(10), 2306 (1006)
7. T. Chiba, J. Ariake, N. Honda, J. Magn. Magn. Mater. **287**, 167 (2005)
8. N. Supper, D.T. Margulies, T. Olson, Y. Ikeda, A. Moser, B. Lengsfield, A. Berger, IEEE Trans. Magn. **42**(10), 2333 (2006)
9. J. Ariake, T. Chiba, N. Honda, IEEE Trans. Magn. **41**, 3142 (2005)
10. G. Choe, A. Roy, Z. Yang, B.R. Acharya, E.N. Abarra, IEEE Trans. Magn. **42**(10), 2327 (2006)
11. J. Chen, H.J. Richter, D.Q. Chen, C. Chang, R. Ranjan, J.L. Lee, IEEE Trans. Magn. **40**(2), 489 (2004)
12. T. Shimatsu, T. Oikawa, Y. Inaba, H. Sato, I. Watanabe, H. Aoi, H. Muraoka, Y. Nakamura, IEEE Trans. Magn. **40**(4), 2461 (2004)
13. U. Kwon, H.S. Jung, M. Kuo, E.M.T. Velu, S.S. Malhotra, W. Jiang, G. Bertero, R. Sinclair, IEEE Trans. Magn. **42**(10), 2330 (2006)
14. H. Nemoto, R. Araki, Y. Hosoe, IEEE Trans. Magn. **42**(10), 2336 (2006)
15. K. Shintaku, IEEE Trans. Magn. **42**(10), 2339 (2006)
16. N. Segikuchi, K. Kawakami, T. Ozue, M. Yamaga, S. Onodera, IEEE Trans. Magn. **41**(10), 3235 (2005)
17. T. Ozue, M. Kondo, Y. Soda, S. Fukuda, S. Onodera, T. Kawana, IEEE Trans. Magn. **38**(1), 136 (2002)
18. J.C. Mallinson, IEEE Trans. Magn. **32**, 599 (1996); H.J. Richter, J. Phys. D: Appl. Phys. **32**, R147 (1999)
19. S.H. Charap, P.L. Lu, Y.J. He, IEEE Trans. Magn. **33**, 978 (1997)
20. Y. Nakatani, Y. Uesaka, N. Hayashi, Jpn. J. Appl. Phys. **28**, 2485 (1989)
21. X. Wang, J.F.d. Castro, K. Gao, Z. Jin, IEEE Trans. Magn. **42**(10), 2294 (2006)
22. D. Weller, A. Moser, IEEE Trans. Magn. **35**, 4423 (1999)

23. K.R. Coffey, T. Thomson, J-U. Thiele, J. Appl. Phys. **93**(10), 8471 (2003)
24. D. Weller, A. Moser, L. Folks, M.E. Best, W. Lee, M.F. Toney, M. Schwickert, J.-U. Thiele, M.F. Doerner, IEEE Trans. Magn. **36**, 10 (2000)
25. U. Nowak, D. Hinzke, J. Appl. Phys. **85**, 4337 (1999)
26. M.H. Kryder, R.W. Gustafson, J. Magn. Magn. Mater. **287**, 449 (2005)
27. J. Fidler, T. Schrefl, J. Phys. D: Appl. Phys. **33**, R135 (2000)
28. T. Schrefl, J. Fidler, J. Magn. Magn. Mater. **111**, 105 (1992)
29. E.C. Stoner, E.P. Wohlfarth, IEEE Trans. Magn. **27**, 3475 (1991)
30. A. Moser, K. Takano, D.T. Margulies, M. Albrecht, Y. Sonobe, Y. Ikeda, S.H. Sun, E.E. Fullerton, J. Phys. D: Appl. Phys. **35**, R157 (2002)
31. Y. Sasaki, N. Usuki, K. Matsuo, M. Kishimoto, IEEE Trans. Magn. **41**(10), 3241 (2005)
32. S. Saito, H. Noguchi, Y. Endo, K. Ejiri, T. Mandai, T. Sugizaki, Fujifilm Res. Develop. **48**, 71 (2003)
33. T. Nagata, T. Harasawa, M. Oyanagi, N. Abe, S. Saito, IEEE Trans. Magn. **42**(10), 2312 (2006)
34. G. Bottoni, IEEE Trans. Magn. **42**(10), 2309 (2006)
35. P.F. Carcia, A.D. Meinhaldt, A. Suna, Appl. Phys. Lett. **47**(2), 178 (1985)
36. J.H. Judy, J. Magn. Magn. Mater. **287**, 16 (2005)
37. E. Miyashita, K. Kuga, R. Taguchi, T. Tamaki, H. Okuda, J. Magn. Magn. Mater. **235**, 413 (2001)
38. S. Saitoh, R. Inaba, A. Kashiwagi, IEEE Trans. Magn. **31**(6), 2859 (1995)
39. A. Matsumoto, Y. Endo, H. Noguchi, IEEE Trans. Magn. **42**(10), 2315 (2006)
40. R.M. Palmer, M.D. Thornley, H. Noguchi, K. Usuki, IEEE Trans. Magn. **42**(10), 2318 (2006)
41. C.T. Rettner, M.E. Best, B.D. Terris, IEEE Trans. Magn. **37**(4), 1649 (2001)
42. G. Herzer, IEEE Trans. Magn. **25**, 3327 (1989)
43. A. Lyberatos, D.V. Berkov, R.W. Chantrell, J. Phys.: Condens. Matter. **5**, 8911 (1993)
44. R. Skomski, J. Phys.: Condens. Matter. **15**, R841 (2003)
45. W. Wernsdorfer, Adv. Chem. Phys. **118**, 99 (2001)
46. G.F. Hughes, IEEE Trans. Magn. **36**(2), 521 (2000)
47. T. Schrefl, J. Fidler, K.J. Kirk, J.N. Chapman, J. Magn. Magn. Mater. **175**, 193 (1997)
48. J. Lohau, A. Moser, C.T. Rettner, M.E. Best, B.D. Terris, Appl. Phys. Lett. **78**, 990 (2001)
49. Z.Z. Bandic, E.A. Dobisz, T.-W. Wu, T.R. Albrecht, Solid State Technol. S7 (2006)
50. M. Ichida, K. Yoshida, IEEE Trans. Magn. **42**(10), 2291 (2006)
51. A.F. Torabi, D. Bai, P. Luo, J. Wang, M. Novid, IEEE Trans. Magn. **42**(10), 2288 (2006)
52. B. Yang, D.R. Fredkin, IEEE Trans. Magn. **43**, 3842 (1998)
53. B. Yang, D.R. Fredkin, J. Appl. Phys. **79**, 5755 (1996)
54. R. Fischer, T. Leineweber, H. Kronmüller, Phys. Rev. B **57**(10), 723 (1998)
55. D. Smith, E. Chunsheng, S. Khizroev, D. Litvinov, IEEE Trans. Magn. **42**(10), 2285 (2006)
56. R.M.H. New, R.F.W. Pease, R.L. White, R.M. Osgood, K. Babcock, J. Appl. Phys. **79**, 5851 (1996)
57. M. Albrecht, S. Ganesan, C.T. Rettner, A. Moser, M.E. Best, R.L. White, B.D. Terris, IEEE Trans. Magn. **39**, 2323 (2003)

58. G.A. Gibson, S. Schultz, J. Appl. Phys. **73**, 4516 (1993)
59. M. Kleiber, F. Kummerlen, M. Lohndorf, A. Wadas, D. Weiss, R. Wiesendanger, Phys. Rev. B **58**, 5563 (1998)
60. C.A. Ross, H.I. Smith, T. Savas, M. Schattenburg, M. Farhoud, M. Hwang, M. Walsh, M.C. Abraham, R.J. Ram, J. Vac. Sci. Technol. B **17**, 3168 (1999)
61. M. Lohndorf, A. Wadas, G. Lutjering, D. Weiss, R. Wiesendanger, Z. Phys. B: Condens. Matter. **101**, 1 (1996)
62. C. Haginoya, K. Koike, Y. Hirayama, J. Yamamoto, M. Ishibashi, O. Kitakami, Y. Shimada, Appl. Phys. Lett. **75**, 3159 (1999)
63. T. Suzuki, Scripta Met. Mat. **33**(10–11), 1609 (1995)
64. T.C. Ulbrich, D. Makarov, G. Hu, I.L. Guhr, D. Suess, T. Schrefl, M. Albrecht, Phys. Rev. Lett. **96**, 077202 (2006)
65. H.J. Richter, A.Y. Dobin, O. Heinonen, K.Z. Gao, R.J.M. v.d. Veerdonk, R.T. Lynch, J. Xue, D. Weller, P. Asselin, M.F. Erden, R.M. Brockie, IEEE Trans. Magn. **42**(10), 2255 (2006)
66. R.L. White, R.M.H. New, R.F.W. Pease, IEEE Trans. Magn. **33**, 990 (1997)
67. S.K. Nair, R.M.H. New, IEEE Trans. Magn. **34**, 1916 (1998)
68. M.M. Aziz, B. Middleton, C.D. Wright, IEE Proc. Sci. Meas. Technol. **150**, 232 (2003)
69. S.Y. Yamamoto, S. Schultz, Appl. Phys. Lett. **69**, 3263 (1996)
70. G.F. Hughes, IEEE Trans. Magn. **35**, 2310 (1999)
71. O. Hellwig, S. Maat, J.B. Kortright, E.E. Fullerton, Phys. Rev. B **65**, 144418 (2002)
72. S. Maat, K. Takano, S.S.P. Parkin, E.E. Fullerton, Phys. Rev. Lett. **87**(8), 087202-1 (2001)
73. B.D. Terris, M. Albrecht, G. Hu, T. Thomson, C.T. Rettner, IEEE Trans. Magn. **41**(10), 2822 (2005)
74. I. Navarro, E. Pulido, P. Crespo, A. Hernando, J. Appl. Phys. **73**, 6525 (1993)
75. Y.K. Takahashi, K. Hono, S. Okamoto, O. Kitakami, J. Appl. Phys. **100**, 074305 (2006)
76. J.W. Harrell, S. Wang, D.E. Nikles, M. Chen, Appl. Phys. Lett. **79**, 4393 (2001)
77. S.H. Sun, C.B. Murray, D. Weller, L. Folks, A. Moser, Science **287**, 1989 (2000)
78. J.A. Christodoulides, M.J. Bonder, Y. Huang, Y. Zhang, S. Stoyanov, G.C. Hadjipanayis, A. Simopoulos, D. Weller, Phys. Rev. B **68**, 054428 (2003)
79. T. Hyeon, Chem. Commun. **8**, 927 (2003)
80. T. Thomson, M.F. Toney, S. Raoux, S.L. Lee, S. Sun, C.B. Murray, B.D. Terris, J. Appl. Phys. **96**, 1197 (2004)
81. H. Kodama, S. Momose, N. Ihara, T. Uzumaki, A. Tanaka, Appl. Phys. Lett. **83**, 5253 (2003)
82. Z.R. Dai, S.H. Sun, Z.L. Wang, Nano Lett. **1**, 443 (2001)
83. Y. Ding, S. Yamamuro, D. Farrell, S.A. Majetich, J. Appl. Phys. **93**, 7411 (2003)
84. A.C.C. Yu, M. Mizuno, Y. Sasaki, M. Inoue, H. Kondo, I. Ohta, D. Djayaprawira, M. Takahashi, Appl. Phys. Lett. **82**, 4352 (2003)
85. S.M. Momose, H. Kodama, N. Ihara, T. Uzumaki, A. Tanaka, Jpn. J. Appl. Phys. **42**, L1252 (2003)
86. T. Thomson, B.D. Terris, M.F. Toney, S. Raoux, J.E.E. Baglin, S.L. Lee, S. Sun, J. Appl. Phys. **95**, 6738 (2004)
87. J.E.E. Baglin, S. Sun, A.J. Kellock, T. Thomson, M.F. Toney, B.D. Terris, C.B. Murray, Mater. Res. Soc. Symp. Proc. **777**, T6.5.1 (2003)

88. H. Sakuma, T. Taniyama, Y. Kitamoto, Y. Yamazaki, H. Nishio, H. Yamamoto, J. Appl. Phys. **95**, 7261 (2004)
89. Y.K. Takahashi, T. Ohkubo, M. Ohnuma, K. Hono, J. Appl. Phys. **93**, 7166 (2003)
90. S.S. Kang, D.E. Nikles, J.W. Harrell, X.W. Wu, J. Magn. Magn. Mater. **266**, 49 (2003)
91. S. Maat, A.J. Kellock, D. Weller, J.E.E. Baglin, Eric E. Fullerton, J. Magn. Magn. Mater. **265**, 1 (2003)
92. B. Vogeli, H.I. Smith, F.J. Castano, S. Haratani, Y. Hao, C.A. Ross, J. Vac. Sci. Technol. B **B19**, 2753 (2001)
93. A. Fernandez, P.J. Bedrossian, S.L. Baker, S.P. Vernon, D.R. Kania, IEEE Trans. Magn. **32**, 4472 (1996)
94. M. Farhoud, J. Ferrera, A.J. Lochtefeld, T.E. Murphy, M.L. Schattenburg, J. Carter, C.A. Ross, H.I. Smith, J. Vac. Sci. Technol. B **17**, 3182 (1999)
95. Y. Hao, F.J. Castaño, C.A. Ross, B. Vögeli, M.E. Walsh, H.I. Smith, J. Appl. Phys. **91**(10), 7989 (2002)
96. A. Carl, S. Kirsch, J. Lohau, H. Weinforth, E.F. Wassermann, IEEE Trans. Magn. **35**, 3106 (1999)
97. W. Wu, B. Cui, X.Y. Sun, W. Zhang, L. Zhuang, L.S. Kong, S.Y. Chou, J. Vac. Sci. Technol. B **16**, 3825 (1998)
98. L. Gao, S.H. Liou, M. Zheng, R. Skomski, M.L. Yan, D.J. Sellmyer, N. Polushkin, J. Appl. Phys. **91**, 7311 (2002)
99. T.A. Savas, M. Farhoud, H.I. Smith, M. Hwang, C.A. Ross, J. Appl. Phys. **85**, 6160 (1999)
100. H.H. Solak, C. David, J. Vac. Sci. Technol. B **21**, 2883 (2003)
101. Y.J. Chen, J.P. Wang, E.W. Soo, L. Wu, T.C. Chong, J. Appl. Phys. **91**(10), 7323 (2002)
102. C.A. Ross, Annu. Rev. Mater. Sci. **31**, 203 (2001)
103. J.C. Lodder, J. Magn. Magn. Mater. **272–276**, 1692 (2004)
104. H.I. Smith, J. Vac. Sci. Technol. B **13**, 2323 (1995)
105. C. Miramond, C. Fermon, F. Rousseaux, D. Decanini, F. Carcenac, J. Magn. Magn. Mater. **165**, 500 (1997)
106. F. Rousseaux, D. Decanini, F. Carcenac, E. Cambril, M.F. Ravet, C. Chappert, N. Bardou, B. Bartenlian, P. Veillet, J. Vac. Sci. Technol. B **13**, 2787 (1995)
107. J.U. Thiele, S. Maat, J.L. Robertson, E.E. Fullerton, IEEE Trans. Magn. **40**(4), 2537 (2004)
108. S. Pramanik, C.-G. Stefanita, S. Bandyopadhyay, J. Nanosci. Nanotechnol. **6**(7), 1973 (2006)
109. R.M. Metzger, V.V. Konovalov, M. Sun, T. Xu, G. Zangari, B. Xu, M. Benakli, W.D. Doyle, IEEE Trans. Magn. **36**, 30 (2000)
110. P. Aranda, J.M. Garcia, J. Magn. Magn. Mater. **249**, 214 (2002)
111. C.-G. Stefanita, F. Yun, M. Namkung, H. Morkoc, S. Bandyopadhyay, *Electrochemical Self-assembly as a Route to Nanodevice Processing*, ed. by H.S. Nalwa. in Handbook of Electrochemical Nanotechnology (American Scientific, Stevenson Ranch, 2006)
112. M. Sun, G. Zangari, M. Shamsuzzoha, R.M. Metzger, Appl. Phys. Lett. **78**, 2964 (2001)
113. C.-G. Stefanita, F. Yun, H. Morkoc, S. Bandyopadhyay, *Self-assembled Quantum Structures Using Porous Alumina on Silicon Substrates*, in Recent Res. Devel. Physics 5: 703, Transworld Research Network 37/661 (2), Trivandrum-695 023, Kerala, India (2004), ISBN 81-7895-126-6

114. C.-G. Stefanita, S. Pramanik, A. Banerjee, M. Sievert, A.A. Baski, S. Bandyopadhyay, J. Cryst. Growth **268**(3–4), 342 (2004)
115. K. Liu, J. Nogues, C. Leighton, H. Masuda, K. Nishio, I.V. Roshchin, I.K. Schuller, Appl. Phys. Lett. **81**, 4434 (2001)
116. R. Zhu, Y.T. Pang, Y.S. Feng, G.H. Fu, Y. Li, L.D. Zhang, Chem. Phys. Lett. **368**, 696 (2003)
117. N.W. Liu, A. Datta, C.Y. Liu, Y.L. Wang, Appl. Phys. Lett. **82**, 1281 (2003)
118. Z.J. Sun, H.K. Kim, Appl. Phys. Lett. **81**, 3458 (2002)
119. S.O. Kim, H.H. Solak, M.P. Stoykovich, N.J. Ferrier, J.J. de Pablo, P.F. Nealey, Nature **424**, 411 (2003)
120. M. Kimura, M.J. Misner, T. Xu, S.H. Kim, T.P. Russell, Langmuir **19**, 9910 (2003)
121. L. Rockford, S.G.J. Mochrie, T.P. Russell, Macromolecules **34**, 1487 (2001)
122. A.A. Zhukov, M.A. Ghanem, A. Goncharov, P.A.J. de Groot, I.S. El-Hallag, P.N. Bartlett, R. Boardman, H. Fangohr, J. Magn. Magn. Mater. **272–276**, 1621 (2004)
123. S. Palacin, P.C. Hidber, J.P. Bourgoin, C. Miramond, C. Fermon, G.M. Whitesides, Chem. Mater. **8**, 1316 (1996)
124. K. Naito, H. Hieda, M. Sakurai, Y. Kamata, K. Asakawa, IEEE Trans. Magn. **38**, 1949 (2002)
125. Z.B. Zhang, D. Gekhtman, M.S. Dresselhaus, J.Y. Ying, Chem. Mater. **11**, 1659 (1999)
126. Z.L. Xiao, Appl. Phys. Lett. **81**, 2869 (2002)
127. B.D. Reiss, C.B. Mao, D.J. Solis, K.S. Ryan, T. Thomson, A.M. Belcher, Nano Lett. **4**, 1127 (2004)
128. E. Mayes, A. Bewick, D. Gleeson, J. Hoinville, R. Jones, O. Kasyutich, A. Nartowski, B. Warne, J. Wiggins, K.K.W. Wong, IEEE Trans. Magn. **39**, 624 (2003)
129. I.W. Hamley, Nanotechnology **14**, R39 (2003)
130. J.Y. Cheng, C.A. Ross, E.L. Thomas, H.I. Smith, G.J. Vancso, Appl. Phys. Lett. **81**, 3657 (2002)
131. C. Harrison, J.A. Dagata, D.H. Adamson, in *Developments in Block Copolymer Science and Technology*, ed. by I.W. Hamley (Wiley, New York, 2003)
132. Y. Kamata, A. Kikitsu, H. Hieda, M. Sakurai, K. Naito, J. Appl. Phys. **95**, 6705 (2004)
133. X.M. Yang, C. Liu, J. Ahner, J. Yu, T. Klemmer, E. Johns, D. Weller, J. Vac. Sci. Technol. B **22**, 31 (2004)
134. J.Y. Cheng, C.A. Ross, V.Z.H. Chan, E.L. Thomas, R.G.H. Lammertink, G.J. Vancso, Adv. Mater. **13**, 1174 (2001)
135. P.W. Nutter, H. Du, V. Vorithitikul, D. Edmundson, E.W. Hill, J.J. Miles, C.D. Wright, IEE Proc. Sci. Meas. Technol. **150**, 227 (2003)
136. D.G. Stinson, S-C. Shin, J. Appl. Phys. **67**(9), 4459 (1990)
137. M. Albrecht, G. Hu, I.L. Guhr, T.C. Ulbrich, J. Boneberg, P. Leiderer, G. Schatz, Nature Mater. **4**, 203 (2005)
138. S.P. Li, W.S. Lew, Y.B. Xu, A. Hirohata, A. Samad, F. Baker, J.A. Bland, Appl. Phys. Lett. **76**, 748 (2000)
139. C. Harrison, D.H. Adamson, Z.D. Cheng, J.M. Sebastian, S. Sethuraman, D.A. Huse, R.A. Register, P.M. Chaikin, Science **290**, 1558 (2000)
140. R.A. Segalman, H. Yokoyama, E.J. Kramer, Adv. Mater. **13**, 1152 (2001)

141. M. Colburn, A. Grot, B.J. Choi, M. Amistoso, T. Bailey, S.V. Sreenivasan, J.G. Ekerdt, C.G. Willson, J. Vac. Sci. Technol. B **19**, 2162 (2001)
142. S.Y. Chou, P.R. Krauss, W. Zhang, L.J. Guo, L. Zhuang, J. Vac. Sci. Technol. B **15**, 2897 (1997)
143. S.Y. Chou, P.R. Krauss, L.S. Kong, J. Appl. Phys. **79**, 6101 (1996)
144. S.Y. Chou, P.R. Krauss, P.J. Renstrom, J. Vac. Sci. Technol. B **14**, 4129 (1996)
145. G.M. McClelland, M.W. Hart, C.T. Rettner, M.E. Best, K.R. Carter, B.D. Terris, Appl. Phys. Lett. **81**, 1483 (2002)
146. M. Natali, A. Lebib, E. Cambril, Y. Chen, I.L. Prejbeanu, K. Ounadjela, J. Vac. Sci. Technol. B **19**, 2779 (2001)
147. S.Y. Chou, Proc. IEEE **85**, 652 (1997)
148. L.S. Kong, L. Zhuang, S.Y. Chou, IEEE Trans. Magn. **33**, 3019 (1997)
149. Y. Soeno, M. Moriya, K. Ito, K. Hattori, A. Kaizu, T. Aoyama, M. Matsuzaki, H. Sakai, IEEE Trans. Magn. **39**, 1967 (2003)
150. N. Tagawa, A. Mori, Microsyst. Technol., Micro-Nanosyst. Inf. Storage Proc. Syst. **9**, 362 (2003)
151. N. Tagawa, T. Hayashi, A. Mori, J. Tribol.-Trans. ASME **123**, 151 (2001)
152. S. Landis, B. Rodmacq, B. Dieny, Phys. Rev. B **62**, 12271 (2000)
153. S. Landis, B. Rodmacq, B. Dieny, B. Dal'Zotto, S. Tedesco, M. Heitzmann, J. Magn. Magn. Mater. **226**, 1708 (2001)
154. G. Hu, T. Thomson, M. Albrecht, M.E. Best, B.D. Terris, C.T. Rettner, S. Raoux, G.M. McClelland, M.W. Hart, J. Appl. Phys. **95**, 7013 (2004)
155. L.S. Kong, L. Zhuang, M.T. Li, B. Cui, S.Y. Chou, Jpn. J. Appl. Phys. **37**, 5973 (1998)
156. J. Moritz, B. Dieny, J.P. Nozieres, S. Landis, A. Lebib, Y. Chen, J. Appl. Phys. **91**, 7314 (2002)
157. J. Moritz, S. Landis, J.C. Toussaint, P. Bayle-Guillemaud, B. Rodmacq, G. Casali, A. Lebib, Y. Chen, J.P. Nozieres, B. Dieny, IEEE Trans. Magn. **38**, 1731 (2002)
158. D. Wachenschwanz, W. Jiang, E. Roddick, A. Homola, P. Dorsey, B. Harper, D. Treves, C. Bajorek, IEEE Trans. Magn. **41**, 670 (2005)
159. C. Ross, Annu. Rev. Mater. Res. **31**, 203 (2001)
160. C.T. Rettner, M.E. Best, B.D. Terris, IEEE Trans. Magn. **37**, 1649 (2001)
161. X.D. Lin, J.G. Zhu, W. Messner, IEEE Trans. Magn. **36**, 2999 (2000)
162. J.G. Zhu, X.D. Lin, L.J. Guan, W. Messner, IEEE Trans. Magn. **36**, 23 (2000)
163. W.H. Bruenger, C. Dzionk, R. Berger, H. Grimm, A. Dietzel, F. Letzkus, R. Springer, Microelectron. Eng. **61–62**, 295 (2002)
164. A. Dietzel, R. Berger, H. Loeschner, E. Platzgummer, G. Stengl, W.H. Bruenger, F. Letzkus, Adv. Mater. **15**, 1152 (2003)
165. D. Ravelosona, T. Devolder, H. Bernas, C. Chappert, V. Mathet, D. Halley, Y. Samson, B. Gilles, A. Marty, IEEE Trans. Magn. **37**, 1643 (2001)
166. C.T. Rettner, S. Anders, T. Thomson, M. Albrecht, Y. Ikeda, M.E. Best, B.D. Terris, IEEE Trans. Magn. **38**, 1725 (2002)
167. C.T. Rettner, S. Anders, J.E.E. Baglin, T. Thomson, B.D. Terris, Appl. Phys. Lett. **80**, 279 (2002)
168. J. Fassbender, D. Ravelosona, A. Samson, J. Phys. D.: Appl. Phys. **37**, R179 (2004)
169. P. Warin, R. Hyndman, J. Glerak, J.N. Chapman, J. Ferre, J.P. Jamet, V. Mathet, C. Chappert, J. Appl. Phys. **90**, 3850 (2001)

170. B.D. Terris, D. Weller, L. Folks, J.E.E. Baglin, A.J. Kellock, H. Rothuizen, P. Vettiger, J. Appl. Phys. **87**, 7004 (2000)
171. T. Devolder, C. Chappert, Y. Chen, E. Cambril, H. Bernas, J.P. Jamet, J. Ferre, Appl. Phys. Lett. **74**, 3383 (1999)
172. J.G. Kusinski, K.M. Krishnan, G. Denbeaux, G. Thomas, B.D. Terris, D. Weller, Appl. Phys. Lett. **79**, 2211 (2001)
173. J. Melngailis, A.A. Mondelli, I.L. Berry, R. Mohondro, J. Vac. Sci. Technol. B **16**, 927 (1998)
174. F. Lu, X.G. Liu, Y. Zhang, V.P. Dravid, C.A. Mirkin, Nano Lett. **3**, 757 (2003)
175. B.D. Terris, L. Folks, D. Weller, J.E.E. Baglin, A.J. Kellock, H. Rothuizen, P. Vettiger, Appl. Phys. Lett. **75**, 403 (1999)
176. S.E. Lambert, I.L. Sanders, A.M. Patlach, M.T. Krounbi, IEEE Trans. Magn. **23**, 3690 (1987)
177. K. Miura, D. Sudo, M. Hashimoto, H. Muraoka, H. Aoi, Y. Nakamura, IEEE Trans. Magn. **42**(10), 2261 (2006)
178. R. Wood, IEEE Trans. Magn. **36**(1), 36 (2000)
179. M. Mallary, A. Torabi, M. Benaki, IEEE Trans. Magn. **38**(4), 1719 (2002)
180. T. Ishida, O. Morita, M. Noda, S. Seko, S. Tanaka, H. Ishioka, IEEE Trans. Fund. Electron. Commun. Comput. Sci. **E76A**, 1161 (1993)
181. H.N. Bertram, *Theory of Magnetic Recording* (Cambridge University Press, Cambridge, 1994)
182. H.J. Richter, IEEE Trans. Magn. **35**, 2790 (1999)
183. D. Weller, M.F. Doerner, Annu. Rev. Mater. Sci. **30**, 611 (2000)
184. K. Watanabe, T. Takeda, K. Okada, H. Takino, IEEE Trans. Magn. **29**, 4030 (1993)
185. S.A. Nikitov, L. Presmanes, P. Tailhades, D.E. Balabanov, J. Magn. Magn. Mater. **241**, 124 (2002)
186. Z.Z. Bandic, H. Xu, Y.M. Hsu, T.R. Albrecht, IEEE Trans. Magn. **39**, 2231 (2003)
187. T. Ishida, K. Miyata, T. Hamada, H. Hashi, Y.S. Ban, K. Taniguchi, A. Saito, IEEE Trans. Magn. **39**, 628 (2003)
188. R. Sugita, O. Saito, T. Muranoi, M. Nishikawa, M. Nagao, J. Appl. Phys. **91**, 8694 (2002)
189. R. Sugita, T. Muranoi, M. Nishikawa, M. Nagao, J. Appl. Phys. **93**, 7008 (2003)
190. T. Ishida, K. Miyata, T. Hamada, K. Tohma, IEEE Trans. Magn. **37**, 1875 (2001)
191. A. Saito, T. Ono, S.J. Takenoiri, IEEE Trans. Magn. **39**, 2234 (2003)
192. http://www.reed-electronics.com/semiconductor
193. M. Plummer et al. (ed.), *The Physics of High Density Recording* (Springer, Berlin, 2001)
194. Y. Tanaka, J. Magn. Magn. Mater. **287**, 468 (2005)
195. S. Gebredingle, S. Gider, R. Wood, IEEE Trans. Magn. IEEE Trans. Magn. **42**(10), 2273 (2006)
196. A. Taratorin, K.B. Klaassen, IEEE Trans. Magn. **42**(2), 157 (2006)
197. A. Taratorin, K.B. Klaassen, IEEE Trans. Magn. **42**(10), 2267 (2006)
198. I.D. Mayergoyz, P. McAvoy, C. Tse, C. Krafft, C. Tseng, IEEE Trans. Magn. **42**(10), 2282 (2006)
199. Z.J. Liu, J.T. Li, K.S. Chai, L. Wang, IEEE Trans. Magn. **42**(10), 2279 (2006)
200. Z. Jin, C. Fu, X. Wang, Y. Zhou, J. F.d.Castro, IEEE Trans. Magn. **42**(10), 2270 (2006)

201. J.R. Childress, M.J. Carey, M.-C. Cyrille, K. Carey, N. Smith, J.A. Katine, T.D. Boone, A.A.G. Driskell-Smith, S. Maat, K. Mackay, C.H. Tsang, IEEE Trans. Magn. **42**(10), 2444 (2006)
202. M. Sekine, T. Watanabe, Y. Tamakawa, T. Shibata, Y. Soda, IEEE Trans. Magn. **42**(10), 2321 (2006)
203. D.W.D. Williams, IEEE Trans. Compon., Hybrids Manuf. Technol. **11**(10), 36 (1988)
204. L. Gomes, *Talking Tech*, The Wall Street Journal (August 22, 2006)
205. E. Grochowski, D.A. Thompson, IEEE Trans. Magn. **30**, 3797 (1994)
206. W. Scholz, S. Batra, IEEE Trans. Magn. **42**(10), 2264 (2006)
207. http://public.itrs.net
208. X.W. Wu, K.Y. Guslienko, R.W. Chantrell, D. Weller, Appl. Phys. Lett. **82**, 3475 (2003)
209. S.Y. Chou, IEEE Trans. Magn. **85**, 652 (1997)

7

Concluding Remarks

Materials research and nanomagnetic phenomena are intertwined fields that inevitably lead to rapid developments of future technologies of increased performance. The interest in nanomagnetics is not just a fashionable trend in today's science and engineering, but rather a natural consequence of technological progress. Novel physical phenomena are uncovered when magnetic systems are reduced to nanoscale.

For instance, semiconductor electronics is traditionally based on the manipulation of charge, nevertheless it is expected that reduction in device dimensions as seen by Moore's law will reach its fundamental limits by 2010 [1]. Since electrons carry both spin and charge, they offer the possibility of combining the two and thereby adding a new degree of freedom to modern devices. An encoder can write information using spin polarization in one part of a spintronic device, while itinerant electrons convey this information to a physically different part where it is decoded. These processes occur in nanodevices, as the spin diffusion length is in that length scale.

Similarly, perpendicular recording media and the challenges posed in magnetic recording by superparamagnetism have their origin in the size of magnetic domains or the width of the domain walls, as well as exchange length, all of them governed by a balance of energy terms at the nanoscale. Nevertheless, magnetism at small length scales has developed in recent years to encompass a wealth of information, solving many of the problems once thought insurmountable.

The most intriguing discoveries in magnetics are yet to come. Depending on future developments, new directions in the many aspects of magnetism will be uncovered, adding to the complexity of this field. These will shape our preferences for certain areas, perhaps rendering others obsolete. Be that as it may, the magnetism field itself can only be enriched as the human mind strives to conquer new realms of knowledge.

Reference

1. H.S.P. Wong, D.J. Frank, P.M. Solomon, C.H.J. Wann, J.J. Welser, Proc. IEEE **87**, 537 (1999)

Index

Springer Series in
MATERIALS SCIENCE

Editors: R. Hull R. M. Osgood, Jr. J. Parisi H. Warlimont

Springer Series in
MATERIALS SCIENCE

Editors: R. Hull R. M. Osgood, Jr. J. Parisi H. Warlimont

74 **Plastic Deformation in Nanocrystalline Materials**
By M.Yu. Gutkin and I.A. Ovid'ko

75 **Wafer Bonding**
Applications and Technology
Editors: M. Alexe and U. Gösele

76 **Spirally Anisotropic Composites**
By G.E. Freger, V.N. Kestelman, and D.G. Freger

77 **Impurities Confined in Quantum Structures**
By P.O. Holtz and Q.X. Zhao

78 **Macromolecular Nanostructured Materials**
Editors: N. Ueyama and A. Harada

79 **Magnetism and Structure in Functional Materials**
Editors: A. Planes, L. Mañosa, and A. Saxena

80 **Micro- and Macro-Properties of Solids**
Thermal, Mechanical and Dielectric Properties
By D.B. Sirdeshmukh, L. Sirdeshmukh, and K.G. Subhadra

81 **Metallopolymer Nanocomposites**
By A.D. Pomogailo and V.N. Kestelman

82 **Plastics for Corrosion Inhibition**
By V.A. Goldade, L.S. Pinchuk, A.V. Makarevich and V.N. Kestelman

83 **Spectroscopic Properties of Rare Earths in Optical Materials**
Editors: G. Liu and B. Jacquier

84 **Hartree–Fock–Slater Method for Materials Science**
The DV–X Alpha Method for Design and Characterization of Materials
Editors: H. Adachi, T. Mukoyama, and J. Kawai

85 **Lifetime Spectroscopy**
A Method of Defect Characterization in Silicon for Photovoltaic Applications
By S. Rein

86 **Wide-Gap Chalcopyrites**
Editors: S. Siebentritt and U. Rau

87 **Micro- and Nanostructured Glasses**
By D. Hülsenberg and A. Harnisch

88 **Introduction to Wave Scattering, Localization and Mesoscopic Phenomena**
By P. Sheng

89 **Magneto-Science**
Magnetic Field Effects on Materials: Fundamentals and Applications
Editors: M. Yamaguchi and Y. Tanimoto

90 **Internal Friction in Metallic Materials**
A Handbook
By M.S. Blanter, I.S. Golovin, H. Neuhäuser, and H.-R. Sinning

91 **Time-dependent Mechanical Properties of Solid Bodies**
By W. Gräfe

92 **Solder Joint Technology**
Materials, Properties, and Reliability
By K.-N. Tu

93 **Materials for Tomorrow**
Theory, Experiments and Modelling
Editors: S. Gemming, M. Schreiber and J.-B. Suck

94 **Magnetic Nanostructures**
Editors: B. Aktas, L. Tagirov, and F. Mikailov

95 **Nanocrystals and Their Mesoscopic Organization**
By C.N.R. Rao, P.J. Thomas and G.U. Kulkarni

96 **Gallium Nitride Electronics**
By R. Quay

97 **Multifunctional Barriers for Flexible Structure**
Textile, Leather and Paper
Editors: S. Duquesne, C. Magniez, and G. Camino

98 **Physics of Negative Refraction and Negative Index Materials**
Optical and Electronic Aspects and Diversified Approaches
Editors: C.M. Krowne and Y. Zhang

Printing: Krips bv, Meppel, The Netherlands
Binding: Stürtz, Würzburg, Germany